THE CITY

URBAN GROWTH

URBAN ECONOMICS

URBAN GROWTH

An Approach

BRIAN T. ROBSON

Routledge
Taylor & Francis Group
LONDON AND NEW YORK

First published in 1973

This edition published in 2007
Routledge
2 Park Square, Milton Park, Abingdon, Oxon, OX14 4RN
Simultaneously published in the USA and Canada by Routledge
711 Third Avenue, New York, NY 10017
Routledge is an imprint of Taylor & Francis Group, an informa business

Transferred to Digital Printing 2007

First issued in paperback 2013

British Library Cataloguing in Publication Data
A CIP catalogue record for this book
is available from the British Library

Urban Growth

ISBN13: 978-0-415-41778-5 (volume hbk)
ISBN13: 978-0-415-41932-1 (subset)
ISBN13: 978-0-415-41318-3 (set)
ISBN13: 978-0-415-85186-2(volume pbk)

Routledge Library Editions: The City

Urban growth: an approach

BRIAN T. ROBSON

Lecturer in Geography, University of Cambridge
Fellow of Fitzwilliam College, Cambridge

METHUEN & CO LTD

First published 1973 by Methuen & Co Ltd, 11 New Fetter Lane,
London EC4P 4EE

Printed in Great Britain by William Clowes & Sons Limited,
London, Colchester and Beccles

ISBN 0 416 66950 6 Hardback
0 416 78710 X Paperback

To Glenna,
Mark and Peter

Contents

Tables

Figures

Preamble

> My centre is giving way, my right is in retreat; situation excellent. I shall attack.
> MARÉCHAL FOCH

Students of the city seem always to be one jump behind reality. The roundabout designed in the '40s is built in the '60s; multiple-storey car parks arise in the centre as multiple shopping centres begin to expand in the outskirts of towns; the form and the functioning of the city change so rapidly and are so little understood that our gropings to tackle problems often produce solutions to the reality of the past rather than the present. For more conceptual problems, the time lag is even greater. The writings of sociologists and anthropologists in the early twentieth century focused on a rural-urban dichotomy at a time when such a distinction had become largely artificial; the human ecologists derived a theory of the city in the 1920s which was outdated almost at the time of its writing; geographers continue to elaborate concepts based on distance constraints when increasing mobility has begun to erode the effects of spatial friction. Much of the theory of urban geography is predicated on a view of cities which no longer exist. They are not unitary, self-contained, definable entities, since the strands of control, influence and effect pass casually beyond the bounds of administrative areas or densely built-up zones. Such conceptual lags are inevitable. Time is needed to refine, to verify and to explore the conceptual frameworks in which theory is couched and it is only on the foundation of some securely-based notions about the recent past that more intelligent guesses about the present and future can be made. One need feel no necessary embarrassment in talking of the 'industrial city' when we are already far advanced in the post-industrial age, since the tasks and the interests of the academic are different from those of the planner, the politician or the futurologist.

Many urban theories make the necessary assumption that the city can be defined as a unit and the fact that such an assumption was more clearly met in the past than the present suggests the value of looking at earlier rather than contemporary data as material to test the concepts of present-day notions. Here, the cities of nineteenth-century England and Wales are studied to examine the assumption that urban population growth obeys a law of proportionate effect – a law which states that, in a set of cities, there is a common rate of growth irrespective of city size, and that individual cities depart from this general rate by random rather than systematic amounts. The approach is an aggregate one; concentrating on the characteristics of the whole set of places rather than upon individual towns. It differs in consequence from much of the writing of urban history which has tended to study the events, the stimuli and the checks experienced, in some unique combination, by only one or a very small number of towns. The two approaches complement each other rather than vie as alternatives; the results of the present study demonstrate the general tendencies inherent in the demographic history of towns while at the same time suggesting the wide bounds within which individual places may grow more or less rapidly than other places of comparable size. The unique events of history add the detail which helps to explain this individual variance, but the overall aggregate curve of growth sets limits to this variation; for the small towns there is considerable scope for individual variation, while for the large town there is considerably less.

At the outset, it was hoped to develop a formal system-based approach to urban growth: regularities in city sizes and in the nature of the overall growth process being seen as symptoms of the inter-relatedness of the individual towns in an urban system; the aim being to isolate the negative and positive feedback loops by which the system achieves stability or progressively alters over time. That this aim was gradually abandoned in favour of a less ambitious and more pragmatic approach, is salutory. It suggests both the limitations imposed by the nature of much of the available data and also the dubiousness of many of the wider claims that the concepts of systems can be applied as working tools in the analysis of social phenomena. The approach is, however, more theoretical than empirical. Even though it draws on the experience and the material of the nineteenth century, its basic concerns are more with the underpinnings of urban theory than with the nineteenth century *per se*. As such it is hoped that it might provide a perspective against which the detailed histories of individual cities might be placed, as well as throwing some new light on the more general questions of urban growth and the diffusion of innovations.

In addition to the specific acknowledgements made in the body of the text, I must thank those people who contributed so much to the preparation of this book. In particular, I owe a considerable debt to Miss Helen Gibson for her valiant and intelligent typing of the manuscript and to Mr Michael Young and Mr William Kirkland for the skill with which they prepared the illustrations.

PART 1

Introduction

CHAPTER 1

Defining the city

> I have therefore looked at the subject without reference to the rivalries of particular towns, and must be understood to admit that in the absence of scientific boundaries comparisons of one place with another can seldom be altogether just.
>
> T. A. WELTON

Ever since the town has been studied there has been agonized debate about its definition.[1] Much of this debate has been contributed by those concerned with the origins of urbanization who, in trying to establish the points in time at which towns first emerged, have needed to clarify with some precision the ways in which 'towns' can be distinguished from objects which are not 'towns' and have therefore needed to develop unambiguous criteria to define their objects of study. It is from this type of concern that the various definitions based on legal (Maine 1861; Maitland 1898), economic (Pirenne 1925) and sociological (Childe 1936; M. Weber 1958) elements have been added to the emphasis on the purely physical aspect of the town as a densely built-up area. In studying more recent periods, the problems of definition have, if anything, become less easy to resolve. Once the legal forms of urban government were codified, once the division of labour and the development of specialized occupations had become fairly widespread, once, concomitantly, the long-distance traffic in men, messages and materials had grown to considerable proportions, and once the built-up area itself had grown to a relatively large size, the distinction between town and country may have looked reasonably precise, but it was not long to remain so. Today, in the developed world, one still thinks and talks of 'towns' as distinct from 'the country', but if it is city air that makes men free who is not a free man in the western world?

By way of introduction, this chapter will attempt to go a little way in clearing the ground for a study of towns by looking at the problems of their definition both in a theoretical and in a more purely operational fashion. It will, however, be more than mere introduction since how we define towns, and how in consequence we measure them, will have great effect upon the studies upon which we might embark and the conclusions which might be reached.

1 The terms 'town' and 'city' will be used interchangeably without any difference being implied.

THE TOWN AS AN OBJECT

In the literature of philosophy and, to a growing extent, in that of geography there is a concern about what is infelicitously called 'entitation'. Any attempt to establish or explore relationships needs first to specify the objects which are to be related in some appropriate manner; and the test of whether the definition is appropriate is that it should be meaningful in terms of the function and the performance of the objects. If a study is concerned simply with description, the stringency of this need is relaxed, but where the concern is with function then it becomes of some moment. Since we shall be concerned with the functions of towns and the relationships between them, the requirements of definition become somewhat taxing and a discussion of the difficulty of thinking of towns as functional objects is not amiss.

Philosophers have suggested a number of types of object: structurally integrated objects, which have volume and whose parts are physically connected; areal aggregates, which are collections of elements physically close to one another but not necessarily structurally bound together; and non-areal aggregates, which are collections defined by some sort of similarity which warrants their being thought of as a 'class'. A beech tree would illustrate the first of these objects, a wood the second and the concept of 'tree' the third. G. P. Chapman (1971) follows Fiebleman (1954) in suggesting that the most meaningful object, and the only one which can be used where functional performance is of interest, is what he calls the 'first-order object of study'. This is an object which can be specified in terms of its morphology, taxonomy, composition, physiology, ecology, chorology and chronology; an object, in other words, which has shape, which is composed of a set of identifiable elements determining its internal functioning, which can be related to other objects defined at a similar scale and having relationships with it at the next-higher scale level, and which varies both over space and time. Such a first-order object of study, meeting these strict requirements of definition, is something which can 'perform' and which can be thought of as a unit at one scale level and part of some more inclusive unit at a higher level of scale. The difficulty, of course, is that such objects can only be defined with any confidence at the extreme ends of the continuum of scale. It is relatively easy to define the successively higher-scale objects of atom, molecule, cell, organ and organism; conceptually it is relatively easy to define the world or the solar system; but it is in the middle ranges of scales where definition becomes difficult and where the blurring of first-order objects and areal aggregates occurs. Geographers have dealt with a multitude of objects at a variety of scales, but the majority of objects with which they are concerned falls right in the middle of the disputed ground between what can and cannot readily be regarded as first-order objects of study. The suggestion that they are concerned with objects above the scale of the atom and below the scale of the universe (Chorley and Kennedy 1971, p. 4) must be regarded as a light-hearted intrusion upon geographers' notorious academic licence, but certainly much of their concern is with objects which are defined in terms of spatial proximity rather than any necessary structural bonding between the component parts – the 'region' being the classic instance. The same, of course, is true of the town. The classic definition of urbanization is provided by Tisdale (1942) who suggests that urbanization is the process whereby increasing proportions of a nation's population are drawn into areas which, being relatively densely settled and relatively large, can be thought of as cities. Here there are two objects involved – the nation and the city – both of which are areal aggregates, but neither of which could be claimed with confidence to meet Fiebleman's strict definition of a first-order object of study. The town, conceived of as an area of densely populated land, may or may not function as a unit, but it would be a brave man who would suggest what exactly it is that functions and what is its sphere of operations. Seen in terms of economic activity, many of the inhabitants of such a built-up area may not work within it and many of the people who work within it would certainly live at considerable distances from it. Likewise, other activities (or performances) of the town as an areal aggregate may affect and involve people from outside its spatially-defined boundary. Such activities would include shopping, recreation, entertainment, the making and execution of a whole variety of political and economic decisions and a host of formal and informal contacts between people and activities.

This is simply to emphasize that, while one can recognize discrete areas which are relatively compact and densely populated, such areas do not necessarily operate in functional terms as units or solely within their own confines. It would seem difficult, if not impossible, to define an unambiguous first-order object to describe 'the town'. Of course this is not an uncommon difficulty; it afflicts all disciplines dealing with objects of middle-order scale. The botanist, for example, is faced with the same problem in trying to define a plant community in terms which would meet the rigour of Fiebleman's definition of an object and many of the suggestions derived from botany bear close resemblance to the functional regions defined by the geographer. For the town, such functional regions have been demarcated in terms of spheres of influence (Dickinson 1947; Smailes 1947; Carter 1955), but the conclusion which arises is not surprising; that no one town has a single sphere of influence, but rather has a set of indistinct spheres each of which differs in varying degrees from the other depending on the particular activity which is used as an indicator of the town's influence on its surrounding area. The tributary area from which a baker draws his custom is very different in extent from that from which a specialist jeweller draws his custom. In terms of the ascending scales of objects, it might be possible to define a baker's shop at a low spatial scale for the hours while the shop is functioning and, at a higher spatial scale, it may again be possible to define a functional object comprising the shop and the households served by it. But when one aggregates together all of the bakers' shops and the jewellers' shops and the multitude of other activities which take place within the town, one has so jumbled a set of low-scale objects comprising the town that any definition of a single functional area which describes the operational extent of the town is faced with so complex a collage that only heroic Occam's-razor decisions could hope to provide any resolution.

One is therefore faced with the problem at the outset that any definition of the city is of necessity bound to be far from satisfactory as a description of the varied types of effect and contact that towns have. Probably in no case could one hope to define a town in terms which would satisfy the requirements of a first-order object of study. But for what objects could one draw such definitions except those of organisms and scales below this?

OPERATIONAL DEFINITIONS

It is against this seemingly intractible conceptual background that the various attempts to resolve operational definitions have been made. The most monumental and conscientious of such attempts have come from America where the Census Bureau distinguishes between such varied definitions as the *Standard Metropolitan Statistical Area* and the *urbanized area* (U.S. Bureau of the Budget 1967), and where Kingsley Davis and his associates at Berkeley have applied common definitions to towns throughout the world (International Urban Research 1959; see also Gibbs 1961). These attempts have tried largely to demarcate towns as built-up areas and they draw attention to the evident unsatisfactoriness of using administrative areas based on political or governmental boundaries. The two main criteria which most often are used in place of political boundaries are, first, some critical population size and, second, some critical level of settlement density, by means of a combination of which villages or non-urban settlements might be distinguished from 'towns'. Such discussion is of great value in making comparisons between sets of national statistics on urbanization, but it obviously cannot resolve the more fundamental problem of the inappropriateness of purely physical definitions of urban areas.

The difficulty has become increasingly severe as the process of urbanization-industrialization has progressed and, with it, the increasing mobility of populations both urban and rural. Increasing physical mobility, allied with the spread of the means of non-physical communication, have tended to pull apart the tightly-bounded town as it has conventionally been thought of. The telephone now enables people to make long-distance contact fairly effortlessly; radio and television disseminate the culture of cities almost universally and instantaneously; and the projection of three-dimensional images by holography (Berry 1970, p. 49) holds out the awesome prospect of the 'embodied' contact of people taking place without their physical movement. Such developments, with the more immediately evident effects of the greater ease and speed of physical movement, have meant that distance has increasingly become less of what Warntz (1967) called a constraining tyranny and, as its power has lessened, so have the tight compaction of cities and the relatively narrow compass of the areas which they affect and which are affected by them, broken down.

A useful distinction which clarifies this effect is that between urbanization on the one hand and the urban process on the other (Popenhoe 1965; Friedmann and Miller 1965; Friedmann 1966b). Reference has already been made to Tisdale's definition of the former as a process by which increasing proportions of people are drawn into cities. One can define the urban process, on the other hand, as that by which increasing proportions of people, irrespective of whether they actually live in cities, are *involved* in ways of life, modes of thought and types of activity which are somehow 'urban' rather than 'rural'. At a time when the tyranny of distance was considerably greater than now, the city's effect was largely restricted to areas within its own spatial extent. Only those who lived within its walls were markedly subject to urban life-styles and were caught up in social and economic activities and types of behaviour which were urban. To this extent urbanization and the urban process could be thought of as roughly synonymous or largely overlapping. Today however, in the most urbanized and economically developed areas, the two are certainly not synonymous. People are sucked into the urban ambit whether or not they live within what we would conventionally define as a town in terms of its built-up area. Indeed, with the general tendency for the populations of large towns to decline where one uses relatively narrow boundaries to define those towns, it might even be said that urbanization is diminishing whereas the urban process gallops on apace.

It is this substitution of an 'urban field' in place of the more familiar 'urban area' that underlies Berry's attempts to define broad swathes of territory and population which are tributary to central cities in the United States (Berry 1968). His definition is based on flows of journeys-to-work and he shows that the streams of commuting produce city areas which overlap and interpenetrate each other in a complex jigsaw within the more densely settled areas of the manufacturing belt, but which elswhere rather accurately dovetail into one another so as to parcel out much of the whole national territory between the various central cities. Such an exercise is of course subject to the same criticisms that might be levelled at any study based on one of a variety of types of spheres of influence, but it is not without interest that Berry finds that no less than 96 per cent of the total American population lies within the set of urban regions which he produces as against only 67 per cent within the census-defined Standard Metropolitan Statistical Areas. Almost the entire United States population thus lives within areas in which at least some portion of the residents have their work in large urban centres and it is significant that it is within those interstitial areas which fall outside these functional economic urban areas that the highest indices of social and economic ill-health are to be found: higher unemployment, lower levels of formal education and other such indices being found in the few areas which lie outside the enlarged compass of the modern city.

THE EVOLVING MEGALOPOLIS

These progressive changes in the role, scope and form of the town have given rise to a number of well-known urban typologies which are based on their changing salient features. The typologies range from the hypnotic writings of Mumford (1938 and 1961) to the prognostications of H. G. Wells (1902 and nicely reassessed by Berry, 1970) and the work of self-styled 'futurologists' such as M. M. Webber (1963, 1964 and 1968). Whatever their tone or stance – gloomy in the case of Mumford, excited in the case of Wells and bland in the case of Webber – the various writers tend to be in remarkable agreement both in regard to the structure of their typologies as well as to the analysis of the principal agents underlying the changes, which are seen as consequences of technological development and particularly improvements in communication (Boal 1968). Mumford's catalogue of eras – eotechnic, paleotechnic, neotechnic and bio-technic; terms originally coined by Patrick Geddes – are basically tied in to the predominant technologies of successive periods and especially to the prevailing fuel technologies. Webber provides a more sociological stress by emphasizing changes in the types of social relationships between individuals, but such changes are themselves the result of alterations in the means of communication. The substitution of face-to-face contact and the primacy of the local community as a basic social group by the emergence of what he calls 'non-place communities' composed of interest groups, is an indication of the greater range of contacts which have been made possible by the greater ease of both physical and non-physical communication. Again, as far as the physical form of the various types of cities is concerned, the compilers of urban typologies draw heavily on changes in the types of internal and external urban transport as the explanations of change.

The urban typologies suggest that the predominant categories of city would comprise the pre-industrial, the industrial and the post-industrial city, each of which would be characterized by differences both in their physical form as well as in their preponderent types of social and economic activities. The pre-industrial city, which has been categorized most extensively by Sjoberg (1960 and 1965; see also the dissenting views of Vance 1971), is characterized by a tightly compact physical form and by social and economic roles which are not highly specialized. Even where one finds, in the exceptional case, pre-industrial cities of some considerable population size, the constraint of distance still produces a 'walking city' since there is a suggestion that the larger pre-industrial cities tend to be composed of a number of relatively distinct component parts each of which conforms roughly to a standard spatial module set by the possible radius of walking and each of which might be physically separated from the other by internal town walls or distinct quarters.[1] The essence of the physical form of the pre-industrial city is therefore its compactness, so dramatically portrayed by such engravers of late medieval European towns as Hogengerg.

The industrial town, by comparison, reflects the loosening of these spatial bonds by the advent of more efficient inter-urban transport and of the simultaneous and gradual transformation of economic and social roles with the development of large-scale factories. The industrial city therefore shows both a much greater specialization of roles amongst its inhabitants and also a much greater physical spread which breaks down the circular compaction of the pre-industrial city. It may be that the industrial city was still compact since intra- as distinct from inter-urban transport was late in improving, but it sends out major tentacles of growth along the main arteries of movement and also, with improvements in both transport and in the techniques of organization, the upper limit of population size which can function as a single whole becomes greater and enables the appearance of cities of considerable size.

The post-industrial city is, superficially, a very different animal. The process of social and economic specialization has progressed even further in terms of fragmenting the roles which individuals play. And, in terms of its physical form, the bounds of the post-industrial city become quite unrecognizable. Not only do improvements in internal transport make it possible for the interstices of the tentacles of earlier growth to be filled in (this perhaps being a feature of the late-industrial rather than the post-industrial city), but increasing numbers of the inhabitants and activities of the city are no longer found within it. The relaxation of the distance constraint has the effect of emphasizing and encouraging the development of groupings of people and sets of economic and social contacts which no longer depend upon physical proximity, but rather upon like interests or complementarity in terms of the contact activity itself. People, in other words, do not merely or even necessarily form social contacts with neighbours, but develop ties with others in terms of shared interests rather than residential proximity. For the city as a whole, with its much greater potential range of control and influence, the once separate and distinct agglomerations of buildings can no longer be taken as delimiting a single functional city entity. Instead there has emerged a polycentric form or a loose amalgamation of centres into a larger and more complex megalopolis (Gottman 1961; Hall 1966). This physical form – or rather the emergence of this formless form[2] – is best seen in the development of vast 'concentrations' of urban areas

[1] It is interesting to note that Marshall (1968) suggests a similar pattern for nineteenth-century industrial towns such as Blackburn and Bury, each of which contained discrete cells centred around a major workplace.

[2] The grammar used to describe post-industrial cities tends often to resort to such antithetical terms as 'community without propinquity' and is indicative of the fact that the quasi-technical language has not yet caught up with the physical reality of the ways in which towns have developed. Notions of continuous space and Euclidean geometry still permeate the terms and have to be countered by the negatives which suggest their inappropriateness.

in the United States. Planners in that country have coined conglomerate names to describe these emergent 'megalopolitan' areas which are as grotesque to the ear as are the phenomena to our conventional urban notions: 'Boswash' for the area between Boston and Washington; 'Chipitts' for the area between Chicago and Pittsburgh; and 'Sansan' for the San Francisco to San Diego area.

While these typologies seem to point to very similar conclusions, they do not suggest that the periods during which one urban 'type' was superseded by another can be pinpointed with any precision. Instead, they have emphasized the role of particular mechanisms whose operation is considered crucial in the long run in the emergence of a distinctively different urban form. These mechanisms emerged and became dominant only over relatively long spans of time. Moses and Williamson (1967) stress, for example, the importance of the motor lorry in improving the internal movement of manufactured goods and raw materials within the city and thus in freeing factories from central locations and in breaking the grip which the centre of the city for so long had on its constituent parts. This was a gradual rather than a sudden development. Since implicit in many of the typologies is the importance of electrical forms of communication replacing mechanical ones, it might be imagined that one could date the emergence of the spatially-unconstrained post-industrial city to the early part of the twentieth century. Certainly it is not merely a product of the last few decades since the megalopolis has emerged very gradually out of the large metropolitan complexes which existed in the nineteenth century and the distinctiveness of the metropolis was recognized by contemporary writers long before the Second World War. For example, the last stage in the urban typology of Gras (1922) was the development of the metropolitan region defined in terms very similar to the later descriptions of post-industrial cities. The Chicago ecologists too, despite their preoccupation with the internal structure and especially the more central parts of a single 'industrial' city, recognized the larger-scale emergence of a distinctive metropolitan complex (McKenzie 1933). Even earlier than this, Mackinder had talked of the influence of London extending widely throughout southern England in terms which are strongly suggestive of precisely the same sort of phenomenon:

> In a manner all south-eastern England is a single urban community; for steam and electricity are changing our geographical conceptions. A city is no longer an area covered continuously with streets and houses. The wives and children of the merchants, even of the more prosperous of the artisans, live without – beyond green fields – where the men only sleep and pass the Sabbath. The metropolis in its largest meaning includes all the counties for whose inhabitants London is 'Town', whose men do habitual business there, whose women buy and spend there, whose morning paper is printed there, whose standard of thought is determined there. East Anglia and the West of England possess a certain independence by virtue of their comparatively remote

position, but, for various reasons, even they belong effectively to Metropolitan England. Birmingham, in Industrial England, is the nearest independent community, with its own heart-beat, with subject boroughs which call it 'Town', with its own daily newspapers guiding opinion not wholly dictated from London. (Mackinder 1902, p. 258)

The gradual recognition of what are considered to be radically different phenomena poses the question of the extent to which these urban types really are different in kind or merely in degree. Is it the confidence of hindsight and the happy blurring of detail through lack of data which enable writers to categorize the pre-industrial city as being so simple and so unitary? Is the dispersed post-industrial city so very different from the industrial city out of which it emerged? Any resolution of such questions would require both more data than are readily available and a separate monograph. It is however worth briefly posing the question of the extent to which the apparent differences are as much matters of the scale at which data can be collected in urban studies as of more fundamental differences of kind.

It may be that, while cities have grown in physical size, our own conceptual scales have not kept pace and our measuring instruments have remained of constant size. This might be illustrated with the familiar example of the shape of population density curves within cities since the literature on the topic suggests that, conforming to the notion that today's cities are different animals, the density curves have progressively changed over time. The regular downward slope of density gradients as one moves outwards from the city centre has been one of the happy lynch-pins in the geographers' quest for spatial regularities and one which has been widely exploited. Following Colin Clark's (1951) first demonstration of a regular negative-exponential decline, the same regularity was shown for countless other cities and the factors universally advanced as explanation invoked the importance of the central city in terms of the bid-rent functions of residential and other land users for all of whom the central city had positive utility. Subsequently, however, this monolithic certainty began to be frayed as writers began to take greater account of the central density crater caused by non-residential land uses and varieties of higher-order surfaces were used to fit the density data (Tanner 1961; Newling 1966 and 1969). More recent still is the suggestion from Berry (1970) that the idea of a density decline is being turned inside-out since population densities now show signs of increase rather than decrease as one moves away from the central business district. This suggestion would appear to tie in neatly with the notion of a post-industrial city which is different in kind from the industrial city which first gave rise to the classic pattern of density decline. It would also find a parallel in the similar erosion of confidence in the existence of a regular distance decline in land values away from the city centre (see, for example, Yeates 1965). Might it not, however, be the case that such apparent differences in the modern city are the result of the scale at which such patterns are measured? Non-residential land has always existed within cities and, if, as seems not unreasonable, it has tended to be found in more central areas, there must long have been a central crater of population density. The fact that the extent of central non-residential land has increased as the area of the city has increased seems undeniable, but the consequence for making comparable measurements of density in cities of different size is that one ought to adopt a unit of observation which differs depending on the extent of the city. That density appears to increase with distance may therefore be a consequence simply of the greater absolute extent of the central city even though relatively the same pattern of centre and periphery may still hold. Once again, the problem may resolve itself into one of trying to define an appropriate extent of what is meant by 'city' and an appropriate module for the constituent parts of which it is composed. The explanation for density curves suggested by Casetti (1967), which relies on the balance of the diametrically opposed forces of desire for centrality and desire for greater space, could as well account for the regular gradients suggested by Clark as for the 'distance accretion' suggested by Berry. Gulliver in Brobdingnag was shocked at the coarseness of the women's skin. Had he been a Brobdingnagian, and not, as he was to them, a dwarf, the women would have appeared as smooth, and no doubt as titillating, as the regular density curves of small cities do to the geographer.

This long digression might perhaps suggest that, while it has indisputably grown larger, while its inhabitants have more widely specialized roles and while its contacts may be more widespread more complex and less contiguous, the modern city may still be the old writ large and that certain of the apparently radical differences are as much a product of scale as of kind.

ESCAPING THE DILEMMA

Nevertheless, we are still faced with a dilemma in that, however the city is defined, all of the operational definitions still leave unanswered the conceptual problem that a city cannot be isolated as a functional operating unit in spatial terms. If it is narrowly defined, many of the people intimately involved in its activities will be excluded; if it is generously defined, many of the people included will be involved only marginally if at all. This suggests that, if we wish to isolate a set of cities with a reasonable amount of confidence in our spatial definitions, a useful compromise might be to turn to earlier rather than more recent periods of time. The extent of flows of commuting into central cities can provide some useful guidelines of the degree of relative self-containment of cities at different periods, and there is a variety of evidence which suggests the changing extent of the journey-to-work over the last century, which demonstrates the gradual widening of the catchment areas of towns. For the nineteenth century it is true that there is little substantive material since census data on place of work and place of residence have only been provided very recently and there are no substantial surveys of such movements comparable to those of the twentieth century. What evidence there is, however, does paint a picture of very short distances separating the worker from his work. Charles Booth's discussion of the casual labour market in London (Booth 1902-3; see also G. S. Jones 1971) gives some idea of the forces making for uncertain employment of large proportions of the working population of the nineteenth century and of their consequent need for access to centres of employment at a time when regular contracts of work were not the normal means of labour supply. The nature of the hiring process, at least as far as many of the unskilled workers were concerned, was conducive to a strong centralizing of the homes of the poor so that they were accessible to the central places where labour was hired. Warnes (1970) gives some suggestive evidence for the town of Chorley by plotting the places of residence of people in particular occupations and the places where such occupations were carried out. While such information is only suggestive in cases where there is more than a single place in which people of a given occupation could be employed, the patterns for mid-nineteenth-century Chorley do show very strong residential concentrations of certain occupations such as miners and quayside workers around the few places where such work was available. From other scattered sources the picture emerges that journeys-to-work were certainly short on average, but that a minority of workers could travel considerable distances even in the nineteenth century. In North America, for example, Goheen's work on Toronto shows that workers in the legal city could be drawn from wide quarters throughout the area for particular occupations (Goheen 1970). Likewise, Pred's maps of the journey-to-work for New York engravers and for miscellaneous industrial occupations in the first half of the nineteenth century (Pred 1966, pp. 196-215) show a generally tight bunching of homes around places of work, but with a certain small fraction of journeys of over two miles. For Britain, the evidence of the mid-1880s submitted to the Royal Commission on the Housing of the Working Classes suggests that it was only the more skilled and better-paid of the artisans who were able to travel distances of more than a few miles. The effects of improving internal transport were limited to a very small sector of the population until well into the 1890s (Dyos 1953 and 1955; Pollins 1964; Kellett 1969, chapter 11; see also Vance 1967) so that, in the words of Hobsbawm (1964, p. 8), for the unskilled artisan, 'all that lay beyond a tiny circle of personal acquaintances or walking distance was darkness'. In a large metropolis such as London, the nineteenth century was almost entirely a period when the great majority of workers lived close to their place of work and when many of the work-places themselves were highly segregated into distinct occupational districts as Hall's (1962) maps of certain of the trades in furniture and clothing suggest. Even though a minority of people travelled long journeys to work in central cities, the predominant pattern of the century was for urban areas to be relatively self-contained employment areas.

For the contrasting picture of the more recent period, there are much more comprehensive and reliable data. Lawton (1968a) has compared the patterns of journey-to-work in 1921 and 1951 for England and Wales and his maps show clearly the increasing pull of the cities, or conversely the increasing spread away from cities of the homes of those who work there. They also suggest that, even in 1921, considerable numbers of workers lived well outside the city limits of the places in which they worked. The massive surveys of Liepmann (1944) for Britain and of Loewenstein (1965) for America provide further corroboration. Liepmann implies in much of what she writes that the lengthening of the journey to work was, in 1944, still a relatively recent phenomenon in Britain and many of her examples certainly demonstrate that relatively large proportions of workers still lived close to their place of work. A survey of Longbridge car workers in Birmingham in 1937, for example, showed that almost one-half lived less than five miles from the factory. Two surveys within London in 1936 found not dissimilar proportions: workers at Standard Telephones and Cables Ltd, with predominantly male employees, had 60 per cent living within five miles; and the tobacco firm of Carreras, with largely female workers, had almost 90 per cent living within the London postal area (Liepmann 1944, pp. 133-45).

While today the blowing apart of the once discrete and relatively self-contained city, which is suggested by the concept of the post-industrial city, is well reflected in the complex overlapping of work-spheres which cover almost the whole of the national territory of developed economies, the process by which the association between place of work and place of residence has become distended is by no means a new phenomenon. It has been developing for as long as means of communication have made it possible for people to move other than on foot with relative ease and increasing speed. The fact that this has been a gradual process means that, giving emphasis to the built-up area of towns and the labour force drawn into that area, it would be difficult to suggest any period for which one could confidently draw bounds to a spatial definition of a city. The further back in time that one goes, however, the greater can be one's confidence in the boundaries that are used.

NINETEENTH-CENTURY TOWNS AS TESTING GROUNDS OF THEORY

The discussion may appear to have moved some way from the problem of urban definition. Certainly the concern has been more with the operational than with the conceptual problems, but if urban population is defined as that collection of people which is affected by work and economic transactions which are characteristically urban – the secondary and tertiary occupations within the context of a sophisticated market economy – then the economic areas traced out by the flows of journey-to-work go some way towards providing a viable measure. Drawing such areas at the present day would raise a host of problems of non-contiguous space since people 'belonging' to city x might live in city y and people living in villages close to the city might not 'belong' to it whereas others in more distant villages might. Such geometrical problems set interesting challenges to the conventional spatial notions of geographers and, while far from being resolved, reactions to them have begun to sketch out the non-Euclidian geometries which would be necessary to tackle them (Tobler 1963; Harvey 1969). But in the absence of ways in which one can readily identify a city and its population as a discrete object, it would seem that the further back in time one goes, the greater the confidence that can be placed in the meaning of the urban populations which can be derived. Since our concern is with population change in towns, this suggests the nineteenth century as a fruitful period in which to test ideas and examine data. Nineteenth-century towns may not have been self-contained, but they were discrete entities to a far greater extent than the complex and sprawling blotches on the map of the twentieth century. Obviously the larger places, such as London or provincial towns of the standing of Manchester, Liverpool or Birmingham, drew upon and affected large tracts of land and large numbers of people from outside their built-up areas and did so to a greater extent as the century wore on. Writing in 1848, and drawing comparisons with the seventeenth century, Macaulay wrote of London, 'The town did not, as now, fade by imperceptible degrees into the country. No long avenues of villas, embowered in lilacs and laburnums, extended from the great centres of wealth and civilization almost to the boundaries of Middlesex and far into the heart of Kent and Surrey' (quoted in Briggs 1968, p. 730). Even though London was later to become, as Mackinder so presciently noted or as Wells so truly prophesied, a metropolis extending its influence over virtually the whole of the country, it was still the case that the divorce between urbanization on the one hand and the urban process on the other had progressed significantly less far in the nineteenth than it has in the twentieth century.

The nineteenth century may, then, offer something of a useful compromise between the unmanageable complexity of defining urban populations at the present day and the utopian state alleged for the self-contained pre-industrial city of early medieval times. It is also, more obviously, a practicable period with which to deal since the demographic data of the censuses and the great wealth of other less general sources of information fully justify Sir Richard Giffen's contemporary epithet of the period as 'the statistical century'. Cairncross (1949, p. 67) considered that changes in the scale and structure of nineteenth-century society meant that there 'are few better quarries for the theory of economic development than Victorian England'. Certainly, for testing theories there is a vast mine of information provided by the censuses, even though, in the words of the well-known quotation, the stones in that quarry are of little use until they are dug out, shaped and built up in an intelligent fashion. Since our concern will be with testing ideas rather than with problems of policy, we would be better served by turning to an earlier period and looking at the nineteenth century as a context in which to develop hypotheses about the nature of the objects we call towns.

CHAPTER 2

City systems and size distributions

> If a model can be developed which, when suitable parameters are fed in, fits everything from the rise and fall of New England stone masons to hospital management and the decay of the city, we may question whether the general system has not become so general as to become operationally meaningless or philosophically otiose.
>
> TIMES LITERARY SUPPLEMENT, 1971

Concerned as they are with so wide a variety of data and with *collages* of those data, it is not surprising that geographers have turned to the concept of systems in their efforts to tease out the relationships in which they are interested. For American geography the notion of the subject as a study of systems has even had a semi-official stamp of approval in the publication, under the auspices of the Association of American Geographers, of *The science of geography* (National Academy of Sciences 1965; see also Association of American Geographers 1968) which argued, following Ackerman's Presidential Lecture of 1963, that the subject was concerned with the study of physical, political and economic systems with locational analysis as an overall set of tools through which such systems might be understood in their spatial context. Reactions to this suggestion in particular and to the application of systems' notions in general have not been lacking. Thoman (1965) protested at the emphasis given to location and the lack of attention paid to historical geography in the agenda formulated in *The science of geography* and Chisholm flailed at general systems theory with an oddly constructed two-tailed whip; it was at once nothing but 'plain common sense' yet an 'irrelevant distraction' (M. Chisholm 1967, p. 51).

Despite such reactions it is apparent that the notion of 'system' is an attractive one to geography. That great prop of geography, the natural region – as conceived by de la Blache in particular – is a system *par excellence* since the concept underlying the recognition of the region is that the various elements which comprise it are given a unity and a distinctive coherence by the interconnection and interdependence which characterize them. This, at its simplest, is what 'system' is all about. As Chorley and Kennedy (1971, p. 2) suggest, it is essentially a *gestalt* concept in which the relationships between the elements make it greater in sum than the mere addition of the constituents of which it is comprised. As Angyal (1941) notes, the difference between an aggregate and a system is that in the former the parts are added while in the latter they are arranged.

A commonly quoted definition of a system is that by Hall and Fagen (1956) who state that a system is 'a set of objects together with the relationships between the objects and between their attributes'. To define and isolate such systems a number of aspects need to be considered. First is the set of objects themselves and here the appositeness of the discussion of first-order objects of study in the previous chapter is apparent. For middle-order objects like the city the geographer is faced with insurmountable difficulties in trying to define an object called 'the city' in functional terms and these problems become of some importance in considering the effectiveness of the application of systems' ideas to urban studies. However one goes about defining the objects of study, they form the elements of any system. The second essential component is the set of attributes of those elements. The cities, or whatever may be the object of study, must have some definable attributes which can be measured and which will provide some indication of the performance of the object. Third, the relationships between the elements need to be specified. These are the connections between the objects which make the system more than a collection of independent parts. Fourth, it must be possible to adduce the state of the system; that particular conglomeration of values of the attributes of the elements which obtains at some given time. Finally, as an over-riding consideration, one needs to specify the relationship of the system to the environment in which it operates. This raises the twin problems of defining the appropriate environment for the system as well as of determining the degree of open-ness or closure of the system itself. This latter point has raised innumerable bogeys in the various studies which have used a systems' framework. Many have used the assumptions which pertain to closed systems in which the state of the system may be expected, by the second law of thermodynamics, to tend towards some form of maximum entropy.[1] In practice of course, virtually all of the systems which are considered in geography are only in varying degrees of closure so that some form of steady state replaces maximum entropy as the 'goal' towards which the system tends. Drainage basins export water to other basins or to the ocean and in turn import water from the atmosphere and from other basins. A nation's economy is likewise supported by imports and exports of goods, people and ideas from and to areas outside its boundaries. How such lack of complete closure affects the performance of the system in question and of the goals towards which it might be hypothesized to develop, are questions calling for some refined discussion. Certainly the common analogy with the tendency for differences in temperature

[1] For the physical sciences, the concept of entropy derives from the tendency of a system to assume the most random molecular state. Systems, in other words, are likely to become disordered. The reason for the tendency is that random molecular motion more probably produces a disordered than an ordered state. Entropy is the property by which the disorder is described quantitatively so that a more disordered state has a higher entropy than an ordered state. In transferring such concepts to urban studies two problems stand out; how can one measure order and disorder and how can one translate the notion of molecular activity into the context of the 'performance' of the city.

of a gas in an enclosed cylinder to equalize over time, cannot be accepted at face value in the application of systems' concepts to social and economic observations.

The use of such systems' ideas within geography has certainly only begun to face the difficulties of transmuting these concepts into ones appropriate to the data used by geographers. Two of the main stumbling blocks have been the confusion between different approaches to systems and the imprecision of the terminology of systems. On the first, Harvey (1969, pp. 471-80) recognizes three sorts of systems' approach. First is systems' analysis in which the concepts of systems developed in biology, the electrical sciences, cybernetics and other physical sciences have been applied to the social sciences simply as tools of analysis for unearthing the relationships of complex bodies of material. Second is the theory of general systems which is an attempt, developed in the writings of Mesarovic (1964), to explore the syntax of systems. Third, and perhaps most controversial, is general systems' theory which, in the hands of von Bertalanffy (1950) and of Boulding (1956), has attempted to demonstrate the universality of explanatory structures by tracing out the isomorphism between systems comprised of very different types of phenomena and thus to establish the lineaments of a grand general theory cutting across the conventional disciplines. The distinction between the writings of Mesarovic and von Bertalanffy is by no means clear-cut, but the contrast between attempts to develop systems' theory as an interdisciplinary philosophy on the one hand and the use of systems' concepts as working tools of analysis on the other is certainly marked. That many criticisms of 'systems' theory' fail to recognize such differences is an example of the confusion which occurs in the literature. At a more detailed level, it is also abundantly clear that there is great confusion in the use of terminology, even in the literature of the physical sciences. So fundamental a concept as entropy is given widely varying interpretations and in this, as in the interpretation of 'systems' analysis' in general, semantic fashion has outstripped semantic clarity (Wilbanks and Symanski 1968). As Langton (1972) has concluded, the euphoria which has sometimes accompanied the discussion of systems' theory in the geographical literature has often been misplaced or at least not supported by sufficiently clear elaborations of its applicability.

SYSTEMS IN THE STUDY OF TOWNS

Most of the attempts to use the concepts of systems within human geography have, either explicitly or implicitly, turned to the more limited use of systems' analysis rather than to the more wide-reaching theories of general systems. Their success has proved rather variable. There have been two main types of application: either the somewhat vague recourse to the terminology of systems and the use of its concepts to provide frameworks within which data are organized or parallelism between phenomena suggested; or the more precise use of certain measures taken from the literature of systems and applied to a more limited body of data. As instances of the former, Berry (1964) has linked together a number of the recurring generalities in urban geography in an attempt to argue that all can be seen under a single neat umbrella. He argues that, in looking at cities, the regularity of density gradients, the tessellations of spacings and sizes and the overall size distribution of cities, provide systems within systems which can be represented by various sets of accounting equations which embody the relationships between the attributes of the city objects. In similar vein, Olsson (1967) draws together the same sets of regularities in tracing the connections between central-place theory, regularities in distance interaction and the stochastic processes which can be taken as underlying them. The more rigorous applications of systems' ideas however have evolved from extensions of 'social physics' and the gravity concepts applied to human populations by Stewart (Stewart 1947; Stewart and Warntz 1958; Neft 1966). The now considerable body of work which has developed such gravity concepts in the context of entropy-maximization has followed the lead of Wilson (1970) whose work on deriving the most-probable states of certain probability distributions has given birth to a wholly new approach to the analysis of the internal structure of cities. Wilson's work has explored various trip-distributions and particularly the journey-to-work and it has translated the gravity models, which have long underlain much of the geographical interest in movement, into but one of a whole family of models of interaction which have subsequently been used as the building blocks for models both at an urban (for example: Crowther and Echenique 1972; Anthony and Baxter 1971) and a regional (for example: Cripps and Foot 1969) scale.

In these latter applications, the use of systems' concepts has been of a more limited sort, as distinct from the much more all-embracing arguments of Berry and Olsson. As Wilson says, the difference is that entropy is used as a measure of a probability distribution and the object of trying to maximize entropy for any particular distribution has been to assess the most probable state of a system whose elements can be defined with some precision and for which distinct constraints can be specified. In his own work, taking a set of origin and destination flows, he produces a matrix of trips which are constrained such that the sum of the elements (the T_{ij} terms which measure trips from origin to destination) equals both the total origins and the total destinations, and the summed total of expenditure on travel in the system equals the overall amount of capital spent on travel. His measure of maximum entropy then represents the most likely state of the system and can be used, in aggregate terms, to distribute trips within the matrix of origins and destinations.

This brings us back once again to the nub of all attempts to use systems' concepts. If the nature of the system cannot be defined with sufficient precision, the application of measures of system becomes inappropriate and unrevealing or, at the very least, difficult to interpret. If there is difficulty in defining towns in the first instance and even greater difficulty in specifying the nature of the linkages between those 'towns', then the use of systems' concepts in justification for empirical regularities takes on more than a slight air of mysticism. For a limited and clearly defined set of objects and their relationships, measures of entropy may have meaning, but for the 'city' as a whole it is difficult to know what they may be. Such doubt surrounds not only those studies which make more nebulous recourse to the language of systems, but also many of the more formal applications of systems' ideas as well. Both G. P. Chapman (1970) and Curry (1964), for example, use cities as their objects of study in using information theory and systems' analysis in the study of the sizes and spacings of settlements. It is interesting to note the difficulty which Chapman finds in interpreting the quantitative measures which he derives to analyse the patterns of settlement distribution and size and equally to note the difficulty of translating Curry's abstract system-based arguments into an urban context without the invocation of some vague and unspecified teleological forces.[1]

CITY-SIZE DISTRIBUTIONS

One of the most frequently advanced justifications for regarding sets of cities as systems has been the appearance of regularity in the distribution of the population sizes of the component places. The regularity of size distributions has been a source both of interest and of frustration within urban geography for a considerable time. While some have argued that it is an indicator of the system-like nature of cities, others have poured scorn on what has been seen as a mindless exercise in curve fitting. Geographical interest in size distributions owes much to Zipf's *National unity and disunity* (1941; see also Zipf 1949), although, as Rosing (1966) has pointed out, Zipf was by no means the first person to point to the regularity of city sizes; Auerbach in 1913, Lotka in 1924 and Singer in 1936 had each previously drawn attention to it and it had not escaped the notice of the Chicago ecologists of the 1920s (Goodrich 1926). Lotka, a statistician, reached a gloomy conclusion after further study of the application of rank-size curves to other data: 'Frequency distributions such as these have a wide range of applicability to a variety of phenomena, and the mere form of such a distribution throws little or no light on the underlying physical relations' (Lotka 1926, quoted in Rosing 1966). Despite this, Zipf's work gave rise to a great deal of subsequent speculation both about the form of the curves which best described city sizes as well as about the mechanisms which might give rise to them.

The first focus of such enquiry was on the contrast between the size distribution suggested by Zipf and that implicit in the earlier suggestion by

[1] See below, p. 29.

Jefferson (1939) of the existence of 'primate' cities. Zipf's 'rule' states that, if cities are ranked from the largest to the smallest and if P_r is the population of the rth-ranked city and P_1 that of the largest city, then city sizes can be described by the simple expression $P_r = P_1 . r^{-1}$. In other words, the second largest city will have a population one-half that of the largest, the third will have a population one-third that of the largest and the nth-ranked city will be one-nth the size of the largest. The 'rule' is referred to as the rank-size rule since rank multiplied by size will be a constant value (equal, of course, to the size of the largest place). Such a distribution of sizes will obviously plot as a straight line on logarithmic paper and the slope of the curve, given by the exponent of 1·0 to which the denominator is raised, will be 45°. The simple Zipf formula can be made more general (Stewart 1947; Berry 1961) by introducing a variable exponent, q, which will steepen or flatten the slope, but still preserve the linearity of the size distribution. Various ways of expressing this straight-line plot using different values for q are shown in fig. 2.1.

A contrasting plot, which would not be a straight line, would arise from Jefferson's suggestion of urban primacy; 'All over the world it is the Law of the Capitals that the largest city shall be supereminent, and not merely in size, but in national influence' (Jefferson 1939, p. 227). Again, to generalize this, the 'primate' pattern would therefore lead to the expectation that a set of cities would contain one or a small number of very large places with a distinct gap between it and an array of very much smaller places.

It is this suggested dichotomy which has given rise to much empirical classification of the size distributions of national sets of cities and to attempts to relate the different curves to aspects of the social and economic characteristics of the countries concerned. Both Colin Clark (1967) and Berry (1961) have drawn up typologies based on size distributions. Clark plots the cumulative frequency of cities against population on logarithmic paper and distinguishes the following types: a pattern of 'primacy' in which the largest cities are disproportionately large so that the right tail of the distribution flattens out; an 'oligarchic' pattern in which, while the number of primate cities is kept in check, there is an over-representation of cities of middle orders of size; and a 'counter-primate' pattern in which the largest cities are under-represented so that the right tail of the distribution steepens. Examples of these different patterns, recognized in such graphical ways, are shown to apply to a variety of countries and Clark discusses, somewhat cursorily, the reasons why different curves might fit different countries. Berry's distinctions are drawn between 'primate' size distributions on one hand and 'lognormal' on the other, with a variety of transitional types between these two extremes. He in fact sees the various curves as parts of an evolutionary scheme in which 'lognormal' distributions might be expected to evolve from 'primate' ones. Like Clark, Berry suggests reasons for the fit of particular curves to particular countries and we shall return to his explanations later in the discussion.

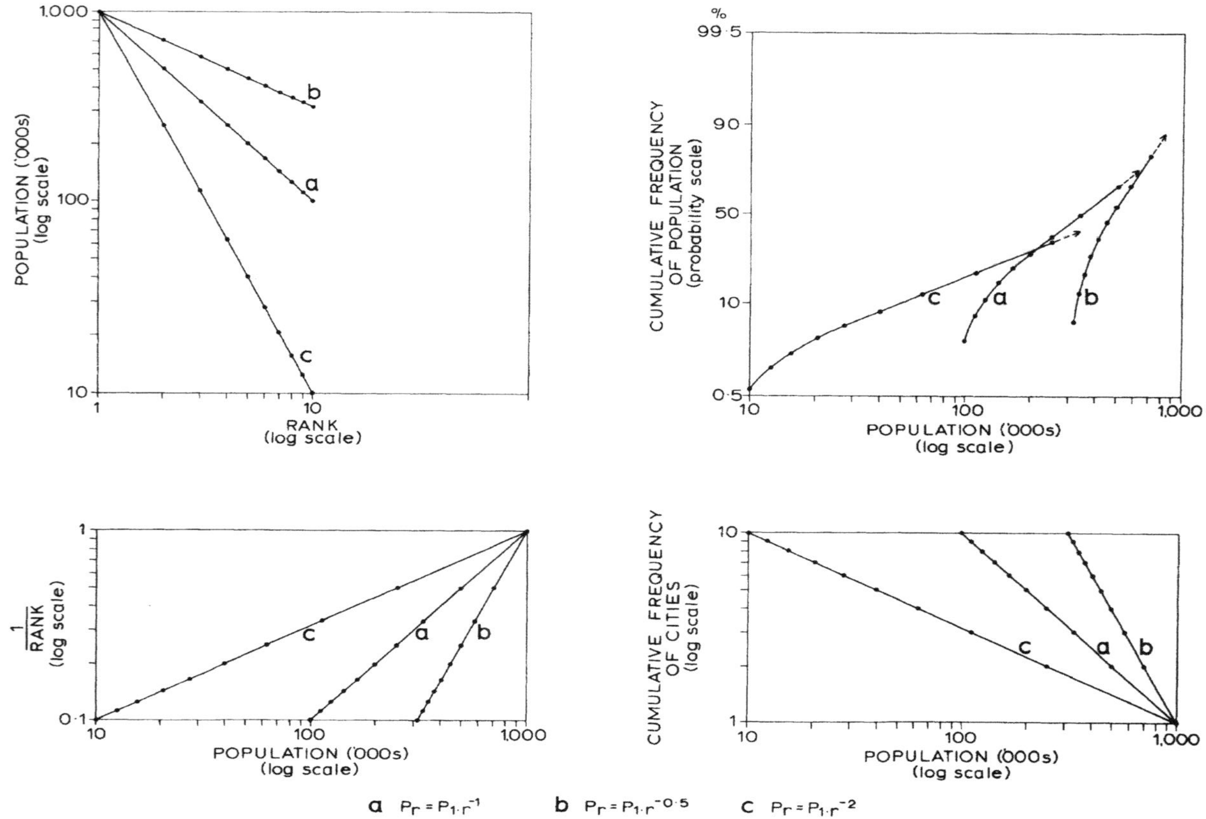

Figure 2.1 The rank–size array: alternative graphical plots. In each diagram the same three size distributions are used

THE VARIETY OF FREQUENCY CURVES

Zipf's rank-size curve is, of course, only one of a number which could be used to describe frequency distributions which are highly skewed and city-size distributions are only one of a variety of types of social and economic phenomena which assume this sort of shape. Many other curves have been suggested and subsequently explored both by statisticians and geographers as predictors of a multitude of phenomena. Zipf himself looked at other data such as the frequency of words used both in speech and in writing and many of the alternative frequency distributions have been fitted to such phenomena as income distribution,[1] population sizes of countries, sizes of firms, the number of publications of scientists and winnings at gambling (for examples of a wide variety of such phenomena, see Kendall 1961). The lognormal distribution has been a popular candidate to describe many such frequencies. Simon (1955) has suggested forms of the Yule distribution and the various forms of Pareto distribution have equally been widely suggested.

The variety of frequency distributions and cumulative frequency distributions which have variously been espoused presents a number of difficulties, chiefly that the distributions are not greatly dissimilar from each other and so are difficult to distinguish. Zipf's law is a higher-the-fewer rule, of which Pareto's equation is a particular case; Zipf's generalized equation is, in turn, an approximation of the much more general Yule distribution. With skewed distributions of the sort used to describe city-size distributions, it is only the right-hand tail which is used to provide a fit to the data and, even though the distributions as a whole may be differently derived, their tails may virtually be indistinguishable given the relatively crude measures of fit which can be called upon. This can clearly be seen in the common assumption that the lognormal and the rank-size distributions are identical, an assumption which Berry makes throughout his work on size distributions. The falsity of this assumption can be seen graphically by comparing identical arrays of city sizes. Figure 2.2 illustrates, for a rank-size array, both a plot of rank against size on logarithmic paper and of cumulative frequency against size on logarithmic probability paper. Where only large cities are used the discrepancy may not be severe, but where smaller city sizes are used, the array of sizes must obviously curve downwards away from the straight-line plot expected on the basis of the assumption of lognormality.[2] Ferguson (1969, Part 2) has provided a formal demonstration of the mathematical lack of equivalence between the rank-size and lognormal distributions. Even though their right tails may be similar (that is, where larger sizes of places are used), the two distributions are not identical. The use of lognormal

[1] Champernowne's (1937) study of income distribution is of particular interest as an early use of transition probabilities before the development of Markov methods.

[2] See also fig. 2.1.

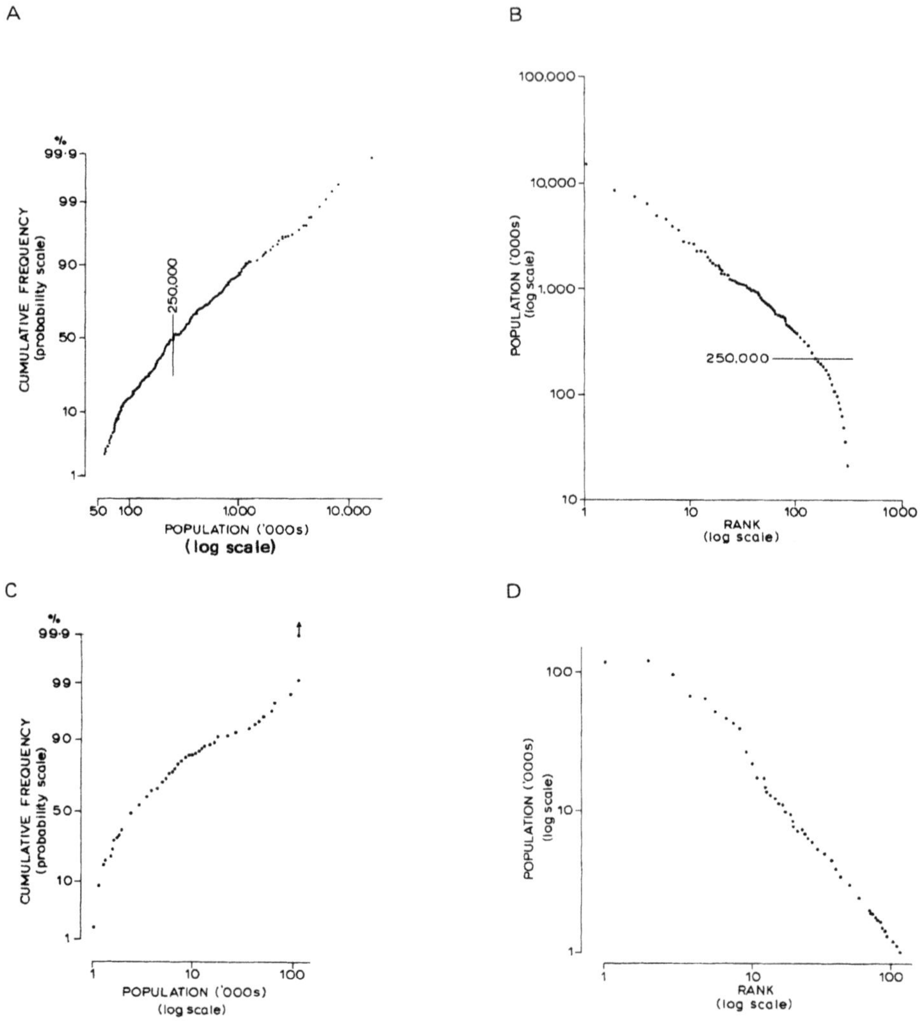

Figure 2.2 The rank-size array: plotting data on logarithmic and log-normal graphs. Data are identical in A and B (Berry 1969) and in C and D (Ferguson 1969)

assumptions to test the rank-size rule is thus fraught with difficulties, amongst which is the choice of a lower population threshold at which places are included since the various graphical portraits of size arrays plotted on logarithmic probability paper produce different slopes, and in certain cases different forms of curve, depending on the threshold size at which cities are selected as qualifying for entry to the city 'system'. In addition, even if the arrays do conform to a lognormal plot, they are not necessarily rank-size curves.

FITTING THE CURVES

Both in fitting and in testing the fit of size distributions there are further difficulties. Tests of the rank-size rule have used a variety of approaches. Stewart (1958), for example, looks at the five largest cities of countries throughout the world and concludes, on the basis of the fact that the ratios of sizes do not conform exactly to the expected progression – 1·0, 0·5, 0·33, 0·25, 0·2 – that the rank-size pattern does not hold. Rosing (1966) also uses only the largest five cities, but fits a curve with a slope of 45° and tests the significance of the chi-square values which indicate the fit between observed and expected sizes. Again he draws a negative conclusion both from this test and also on the basis of the lack of relationship between the residuals from these curves and various measures of the economic development of the countries. Whereas Stewart's method of fitting a curve to the largest place makes the unreasonable assumption that all places other than the largest may be out of kilter, Rosing appears to use the more precise method of least-squares fitting. Browning and Gibbs (1961) provide a rough approximation to this least-squares procedure which involves calculating the expected population of the largest place by summing the reciprocals of all of the city ranks, and then deriving the expected populations of other places as fractions of this in the progression 0·5, 0·33 etc. Such methods all assume that the fit is to be against the simple rank-size plot using an exponent of 1·0, but the most usual method is least-squares regression which produces both a variable intercept and a variable slope to the curve (see Duncan and Reiss 1956, pp. 25-8 for an early example). The goodness of fit can then be tested by using chi-square, although the further difficulty arises here that the density of point observations varies along both axes so that differential weight is given to the large and the small places in the fitting of the curve.

The most extensive test of the various alternative curves is that by Quandt (1964) who examines no fewer than eight types of distribution which are tested against city sizes amongst other types of data. Since the Pareto type-I was fitted by two different methods – using least squares (Pareto type-1a) and the method of maximum likelihood (Pareto type-Ib) – a total of nine tests was used in all. The city-size data were treated in a variety of ways: by using a basic set of cities of over 50,000 population; by using the same set, but with 50,000 subtracted from each observation; and by looking only at the upper half of the distribution. The goodness of fit was tested with the Kolmogorov-Smirnov statistic which, as Quandt admits, is not a wholly satisfactory test since it measures only the overall displacement of the expected and observed curves and cannot compare the two curves along their entire range. Of the distribution functions most commonly invoked in the case of city sizes – the lognormal and the Pareto – it is of considerable interest to see that Quandt found that each proved to provide a best fit to one or other of the various data sets. For example, with the full set of cities of over 50,000 population, all four distributions met the critical value so that it was not possible to reject the null hypothesis that each could have generated the sample set of city sizes; but, of the total of nine tests, the distribution function which performed best was the Pareto type-Ia, second was the Pareto type-Ib, third was the Pareto type-II and fifth was the lognormal. A second approach involved comparisons of the lognormal and the two Pareto functions with the coefficients of the Lorenz curves of the city sizes and, here, it was the Pareto type-II and the lognormal which proved to be the best predictors. Finally, using in place of the Kolmogorov-Smirnov a new statistic which incorporates observations at each point along the whole array of observations, Quandt suggests that the lognormal is a much better description of the modified set of city sizes (with 50,000 subtracted from each population) than for the unmodified set, thus to some extent reinforcing the suggestion made above that the lower population threshold of size which is selected in testing such distributions is critical both in fitting and testing the fit of distribution functions.

Quandt's survey is of interest in a number of ways. First it suggests the difficulty of establishing the goodness of fit of such frequency distributions. Second, and only partly consequent upon the first point, it suggests the even greater difficulty of distinguishing between the fit of various of the alternative distributions. Each is closely similar to the other and, while no single one could explain all the sets of data at which Quandt looks, for any one set of data more than a single distribution will give a reasonable estimate of the actual observations. It is this latter point which provides the most telling statistical weakness of the voluminous literature which argues for one rather than another frequency distribution as an explanation of such socio-economic data as city sizes. Quandt concludes that 'the distributions under consideration are all fairly similar; they probably exhibit smaller deviations from each other than a perfectly error free sample would exhibit from the actual one' (Quandt 1964, p. 22).

There is a close parallel between such curve-fitting exercises applied to city-size distributions and the attempts to fit probability distributions to the frequencies of observations in spatial quadrats which are found in the geographical literature on point-pattern analysis (Harvey 1966 and 1968). In this latter literature, such distributions as the negative binomial and the Neyman type-A have been compared with empirical data on the frequencies of point observations such as settlements. The conclusions which emerge suggest the difficulty, similar to that of city-size distributions, of identifying unambiguously which of several distributions best fit the appropriate data. Furthermore, as Dacey (1968, p. 61) notes, no fewer than six different generating processes might give rise to a distribution such as the negative binomial. Harvey, reflecting on this, has suggested that the point-pattern analyses are examples of what he calls 'over-identified' models since, without further information, it is not possible to infer which of a number of theories might be most appropriate

(Harvey 1969, p. 166). This parallel suggests that one might profitably turn to the process explanations which have been suggested for the various city-size distributions to see to what extent they too might be thought of as over-identified, since, as has been suggested, the critical right-hand tails of these distributions are similar and yet the distributions themselves are severally produced by different types of generating process.

EXPLANATIONS OF CITY-SIZE REGULARITIES

Three general types of explanation of city-size distributions might be recognized: concepts derived from central-place theory; ideas based on principles of maximization or optimization; and notions based on the assumption of the random growth of city populations. The three are by no means clearly distinct. The second and third in particular overlap considerably since random growth is usually seen as a product of the same types of varied and multiple stimuli which lead to the overall entropy-maximizing steady states which underlie many of the second types of explanation. It is useful, however, to discuss the three separately.

The first is the only one which does not allow for any operating dynamic process of growth since central-place theory has never been couched in dynamic terms. A central-place approach is therefore hard to reconcile with a system which grows by small increments as do city populations. There has, however, been a number of attempts to demonstrate the compatability of the carefully balanced geometry of central-place theory, with its expectation of distinct breaks in a hierarchy of centres of different levels, on the one hand, and the regular and continuous size gradation of the rank-size rule on the other. If such a relationship could be demonstrated formally, it could be argued that the spatial rationale of central-place theory – optimal space-filling and distance minimization on the part of consumers – would apply equally to the rank-size rule. The suggestion which has been most fully explored is that, while distinct hierarchical levels do exist, as central-place theory would suggest, the multiplier by which the population size of cities at one hierarchical level is 'blown up' to that of the next level is a random multiplier so that the steps of the size distribution are blurred into a continuous rather than a stepped curve. This attempted resolution has been widely invoked, but its only formal demonstration (Beckmann 1958) is undermined by a number of statistical and operational mistakes which have been pointed out by Dacey (1966) and, more especially, by Parr (1969; see also Beckmann and McPherson 1970). Parr (1970) has provided a review of these attempts to resolve the two size distributions and, while unable to demonstrate the compatability of the two, does show that the rank-size distribution can produce parameters consistent with central-place hierarchies so that, to this extent, the conceptual bases of the two are not substantially dissimilar. There is still however no formal proof of the idea, conceptually appealing though it may be.

Berry (1967, chapter 2) has suggested a slightly different mechanism by which the theory of central places and the empirical observations of rank-size distributions might be reconciled. He suggests that at different regional population densities the threshold sizes of the identical hierarchical levels in a central-place system will differ since, at a lower population density, even though the trade area of a given *m*th-level town will increase, the increase will not be sufficient to keep the supporting tributary population at the same size as that for an area of higher density. Thus, while in a small and relatively equally densely populated area, city sizes might be expected to show distinct hierarchical breaks, when a number of such small areas, each with its own regional population density, are aggregated together the differently-sized hierarchical steps would blur the overall distribution of city sizes into a more truly continuous curve. Again, attractive as the idea may be and despite the demonstration that threshold sizes do vary with regional population densities, there is nowhere a formal demonstration that the blurring would produce a lognormal, a rank-size or any other form of distribution. The distribution would presumably be dependent upon the particular *collage* of areas and densities which were pooled to form the aggregate 'system'. Even if the suggested isomorphism between central-place theory and the rank-size rule could be demonstrated, this first set of explanations would in any case be subject to the same damaging criticisms which have been widely levelled at central-place theory.

The second set of explanations calls rather loosely upon a number of somewhat ill-defined 'principles'. Zipf's own explanation invoked the joint effect of what he saw as two opposing forces: one being a force of 'diversification' by which a large number of small places develop, benefiting from locations close to widely distributed raw materials and thus minimizing their transport costs; the other being a force of 'unification' whereby a small number of large places develop to minimize the transportation costs to the consumer. The emergence of a rank-size distribution reflects the achievement of 'balance' between these opposing tendencies, although it is by no means clear how such balance emerges or why it should lead to so striking a regularity as Zipf was trying to explain. A similar sort of mildly mystical force appears to underlie many of the explanations which have been couched in terms of entropy. When Berry and Garrison, for example, talk of the rank-size distribution being the outcome of a random state and 'as such it is a condition of entropy' (Berry and Garrison 1958, p. 90), it is far from clear what they mean by 'entropy'.

There are, however, more satisfactory explanations invoking maximization or optimization which call upon more formal statements of entropy-maximization. The most precise has been applied not to city sizes, but to word frequencies. Mandelbrot (1953), drawing on information theory, derives a word-frequency distribution in terms of a principle of maximizing information content and, in similar vein, derives a frequency of word lengths in terms of maximizing efficiency or, alternatively, of minimizing cost. The closest parallel to this in the city-size case would seem to be Curry's (1964) rather less precise attempt to derive a city-size distribution. He suggests the following approach, which draws on the notion of random behaviour and the concepts of information theory. If N people are divided between Z settlements, with i being the number of people in each settlement and n being the average number of people per settlement, the cumulative distribution when entropy is maximized will be $P_1(1 - e^{-i/n})$. This, Curry suggests, produces a situation in which the ratio of the population of the m + 1st city to the mth city will be constant. This would obviously give a rank-size pattern, but it is not clear from his exposition how this ratio is derived. Nor, indeed, is it clear how one might translate this notion into processes of city growth although in a later paper Curry (1967, p. 231) appears to suggest that the rank-size pattern results in minimizing the length of the journey to shop in a central-place system.

Another, and very novel, derivation of a theory of city sizes which involves minimization is that of Evans (1972) who views cities as being composed of 'coalitions' of manufacturing firms. Each city's size is a linear function of the number of firms contained within it and, for each firm, economic returns are a function of city size. Each firm will tend to join the coalition of that size which minimizes its production costs and a general equilibrium will be achieved as an n-person game in which each firm, acting independently, joins coalitions of other firms. As an example, assuming that proportional deviations from the optimal coalition size result in equal proportional changes in production costs, Evans derives a model of city sizes in which each city is three times the size of the next largest place. Despite the attractiveness of the general approach of the paper, however, the detailed size distributions which are produced are highly dependent upon the initial assumptions which are made about the numbers of firms for which particular size coalitions are optimal.

A further approach, involving optimization rather than maximization, is represented by explanations which call on the principle of allometric growth in city systems. Stemming from the fascinating work of biologists such as D'Arcy Thompson (1917), allometric growth suggests that there is a constant ratio between the growth rate of an organ and of the organism of which it is a part. The farmer who measures a pig's weight from its girth rather than by heaving it onto a weighing machine, is a practical advocate of a principle of allometry linking width and weight in living organisms. Translating the biological notion into the city case, the proponents of allometric growth argue that, if one plots two curves of city sizes at successive times, the horizontal shifts represent the growth of the organism as a whole (the addition of new cities into the urban system) while the vertical shifts represent growth of the component organs (the changes in the individual city populations). Where the two curves are parallel, as

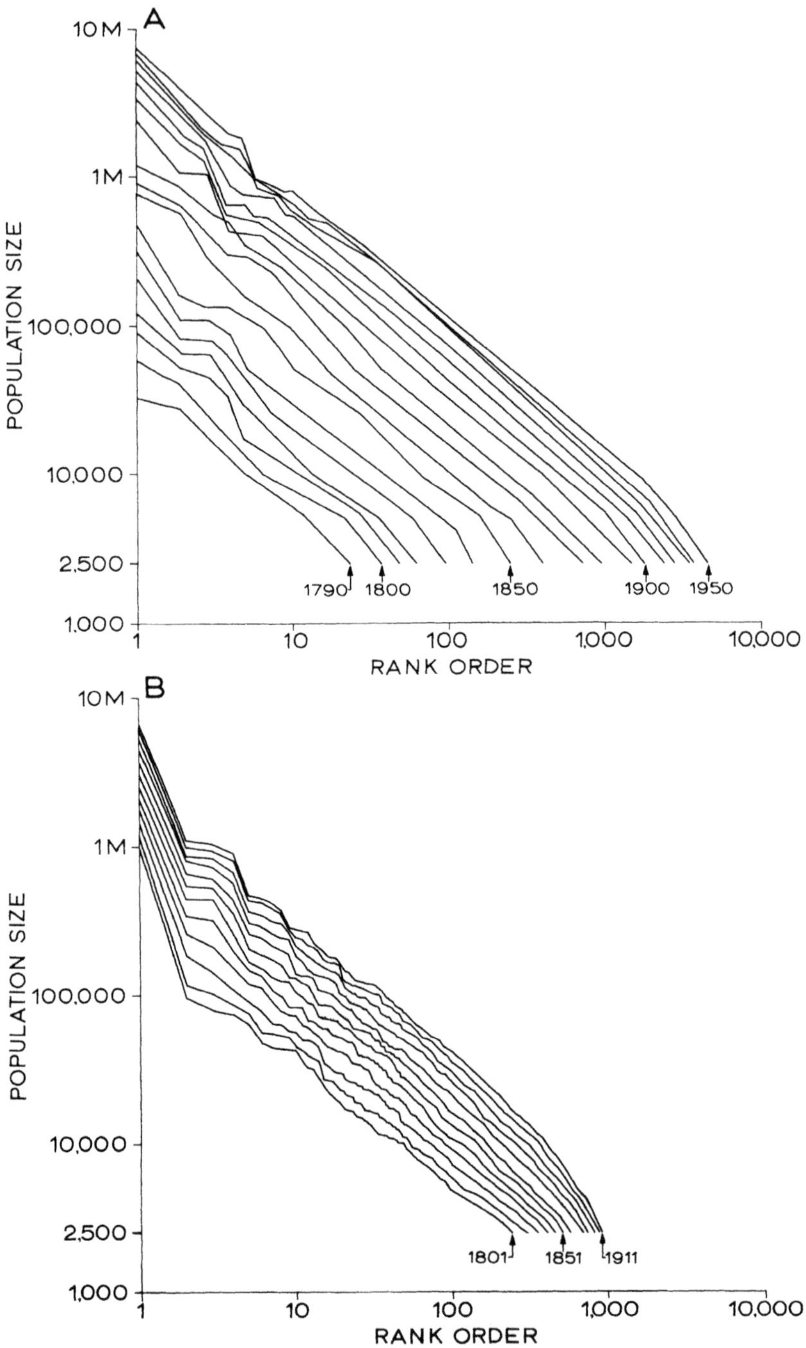

Figure 2.3 The size arrays of cities over time: (A) United States, 1790–1950 (B) England and Wales, 1801–1911

indeed is often the case in empirical data (fig. 2.3), then the ratio between the two is constant, as the law of allometric growth would suggest. Nordbeck (1965 and 1971) first introduced the idea to geography in looking at the constant ratio between city population and area, seeing allometric principles on the basis of a number of logarithmic relationships which he demonstrated both for this and for a number of other relationships applying to social and physical phenomena. Woldenberg (1967) has translated this allometric approach to the arguments surrounding city-size distributions by suggesting that rank-size plots are the consequence of an allometrically-growing central-place system.

The difficulty underlying acceptance of these various arguments which invoke principles of optimization or maximization is how the 'principles' might be translated into real-world conditions. The problem of trying to define an object called a 'city' has already been discussed briefly in relation to the application of systems' notions in general. This basic difficulty makes it extremely problematic as to what it is that the various explanations which call on optimization or maximization are trying to optimize or maximize, since, if one has difficulty in isolating the object itself, it is even more difficult to conceive of that object having some goal towards which it 'operates'. And, given the complexity of the relationships between the component parts of the total city whose population size is being used, it is unrealistic to attempt to single out one component and claim that it is shopping or space or whatever which is the sole element being optimized or maximized. Perhaps it might only be claimed that, by using the expectation of a maximally-probable state in a situation of such complexity that it can be thought of as 'behaving' randomly and thus subject to the laws of probability, measures of maximization can provide operational definitions of unconstrained situations against which departures will appear significant as clues to the types of constraint which operate within the system. This idea that the rank-size or the Pareto or the lognormal forms of city-size distributions may represent ideal unconstrained states and that it is the variations from these ideals which should be of interest as indications of the existence of constraints, is a suggestion which has certainly received much support in the literature on rank-size distributions (for example: Olsson 1967, p. 35; Berry 1964, pp. 149-50; Rosing 1966).

The third set of explanations of the city-size regularities looks explicitly or implicitly at the growth processes which are implied by different types of frequency distribution or which produce such distributions. They suggest that a rationale for the appearance of a given distribution may therefore lie in the reasonableness of the growth assumptions which are used. Two slightly different growth processes have been assumed: the first that, within any particular size group of cities, the towns will grow by a random proportion of their previous population; the second that any individual city will grow by a random proportion of its existing population. The former assumption underlies part of Simon's stochastic model of size distribution. He suggests (Simon 1955) that a

Yule distribution describes a variety of data including the city-size case. His argument for the Yule as a generating process is developed by using a word-frequency illustration. Given S words in a text, the probability that the S + 1st word has already appeared i times is assumed to be proportional to the number of words that have already appeared i times. A second assumption is that new words will appear with a constant probability. The right tail of his cumulative distribution – $f(i) = a \,.\, i^{-c} \,.\, b^i$ – reduces to the frequency equivalent of the rank size when b equals 1·0 and so the second assumption, which can be translated into the city case as being that new cities will be added to the system at a constant rate, becomes important in terms of the closeness of the b parameter to this figure of 1·0. It is Simon's first assumption, however, which deals more directly with the growth of existing cities since it argues that city growth would be random within a given size range of cities; in other words that fresh population would be added to existing cities as a constant, but random, proportion of their existing size. This assumption of random growth within groups of city sizes is a relatively weak one since it allows *individual* city growth to be influenced by the growth rate of previous periods (Steindl 1965; Simon and Bonini 1958; and, especially, Ijiri and Simon 1964). A given individual city which grew rapidly in one time period might, therefore, still be expected to grow rapidly in a subsequent period and, so long as its temporally-autocorrelated high growth rate was compensated for by some other city which grew commensurately slowly, the overall distribution of growth rates within that size class of cities might nevertheless be randomly normal. The assumption could, therefore, readily embrace the case of individual cities which were favourably placed with respect to growth so that an individual city's growth might be 'determined' yet the distribution of the size class as a whole still be random. E. N. Thomas' (1967) growth assumption is more restrictive. In deriving a rationale for a lognormal size-distribution, he assumes that growth will be random for any individual city and will therefore not be influenced by its previous growth rate. Thomas argues that, given this assumption and starting from a set of cities whose population at the outset is normally distributed, after a sufficient number of periods of small incremental growth, the distribution will converge on the lognormal. Looking at empirical data from Iowa, he finds that this lognormal model provides a less good fit than a log-lognormal model which he suggests as an alternative, but the importance of his paper lies in the rationale which it suggests for the lognormal curve which has been so widely used in describing city sizes. It is, of course, the case that most of the writers who have tested for lognormal distributions have equated it and the rank-size distribution so that one rapidly gets into deep water in trying to match the sets of growth processes which they have assumed to the particular frequency distributions which they have tested.

These two mildly different growth assumptions have provided the background to a number of the studies of city-size distributions which have called somewhat vaguely upon the complexity and hence the essential randomness of urban growth rates. Once their article is stripped of its systems' language, this is the argument underlying Berry and Garrison's (1958) explanation of city sizes. Most explicitly, Berry himself has advanced this argument in relation to his work on the relationship of 'lognormal' patterns of city sizes in various countries and their degree of economic development. He argues (Berry 1961 and 1969) that a 'lognormal' pattern can be expected to characterize those countries in which – because of their size, their complexity, their long history of urbanization and their well-developed linkages which bind all of the cities into a single economic nexus – there are many and varied growth stimuli operating on their cities. This is in distinction to countries which are small, or have only begun the process of urbanization, or which have a dual economy with only one set of places being tied in to commercial activity, and in which the growth stimuli might be few and might be restricted to a subset of places so as to produce differential growth between one set of places and another. Essentially, the underlying rationale is that of providing reasons to expect a common growth rate to apply to the whole array of cities, a growth rate which would only be disturbed by random fluctuations about that mean. It is not clear whether it is the Simon or the more restrictive Thomas assumptions which are invoked, but the general argument is for some form of random growth.

More recently, Berry (1969 and 1972) has added to these notions a connection with the growth stimuli which arise from the diffusion of technical innovations throughout a national economy and argues that, where a country is integrated into a single economic system, the spread of such innovations would filter through the whole array of cities in a manner which would lend support to his assumption of random growth. This argument has also been pursued by Ward (1963) in a paper whose argument is worth repeating in some little detail. He argues that many of the models of size distribution require random urban growth and that, while this may be reasonable so far as the purely biological mechanisms of birth and death rates are concerned, much urban growth is the result not of natural population change, but of migration. Indeed, it might be argued that all growth within a city is the product of migration since those people who are born and remain within a particular city can be considered to have made a decision to migrate a zero distance. If one therefore considers all growth to be a product of migration and if that migration is considered to be the product of differential economic opportunities within the cities, then it is the factors affecting the number of economic opportunities within cities towards which attention should be directed. Ward argues that the differential distribution of such opportunities can be considered to be the result both of technological change within a given city and of the expansion of its market opportunities. Both of these can be subsumed under the effects of market expansion in a general way since technological change can only find commercial expression through the willingness of consumers in a market to buy the products of a new technology. He then argues as follows:

> The size of the population of a city is a measure of its market size so that market-expansion opportunities depend for their occurrence on the number of inhabitants a city possesses. The opportunities characterized in this way then occur with a relative frequency (probability) which is proportionate to the size of the market. There is some minimum city size below which the probability of occurrence of opportunities is much smaller partly because there is less effective search by both prospective entrepreneurs and prospective migrants, and partly because the basic natural and technological material for opportunity creation does not exist in any quantity. These assumptions are sufficient to generate the Paretian city size distribution. (Ward 1963, pp. 215-16)

Thus, in terms of a rather more extended argument which ties in the effects of migration and the spread of technological change, Ward suggests – although not in formal terms – a reason why city growth above a certain size of city might be a constant fraction of previous size with some disturbance term added in the form of a distribution of probabilities around this common rate.

SYSTEMS AND CITY SIZES

The discussion has so far looked at the various types of frequency and cumulative frequency distributions which have been suggested as generators of city sizes and at some of the explanations offered in support of the regularities which appear. Behind most of this concern with size distributions has been the belief that regularity betokens the existence of a system of cities. In the light of our earlier criticisms of the application of systems' concepts to such vaguely-defined objects as cities, what conclusions can be reached in respect of the validity of such claims?

A study which has tried to carry the idea of systems of cities to a logical geographical conclusion is Harris' (1970) work on Russian cities and, despite its novelty, it demonstrates that the use of size distributions as an indicator of city systems is a very blunt-edged tool. Harris suggests that, for any given territory, the emergence of a regular rank-size array might suggest the existence of a functional system. Within a national area, it might happen that a number of sub-systems would be found, reflecting the lack of overall connectedness of the national economy. These system and sub-system characteristics might, he suggests, be isolated by studying the size distributions of the component cities. Supposing, for example, that a rank-size plot betokened the existence of a functional system and that, in a given area, one had three such plots which reflected the existence of sub-systems within the whole area, then the superimposition of their respective size distributions would lead to a curve of the sort traced out in fig. 2.4 by the line J1, J2, J3, K4, K5, K6, L7, L8, L9 . . . S28, S29, S30. If ten such sub-systems were included, the curve would be traced by the line, J1 . . . J10, K11 . . . K20 . . . S91 . . . S100. In each case, while the lower end of the curve would increasingly approximate the rank-size curve, the upper end would be marked by noticeable breaks above which the slope would

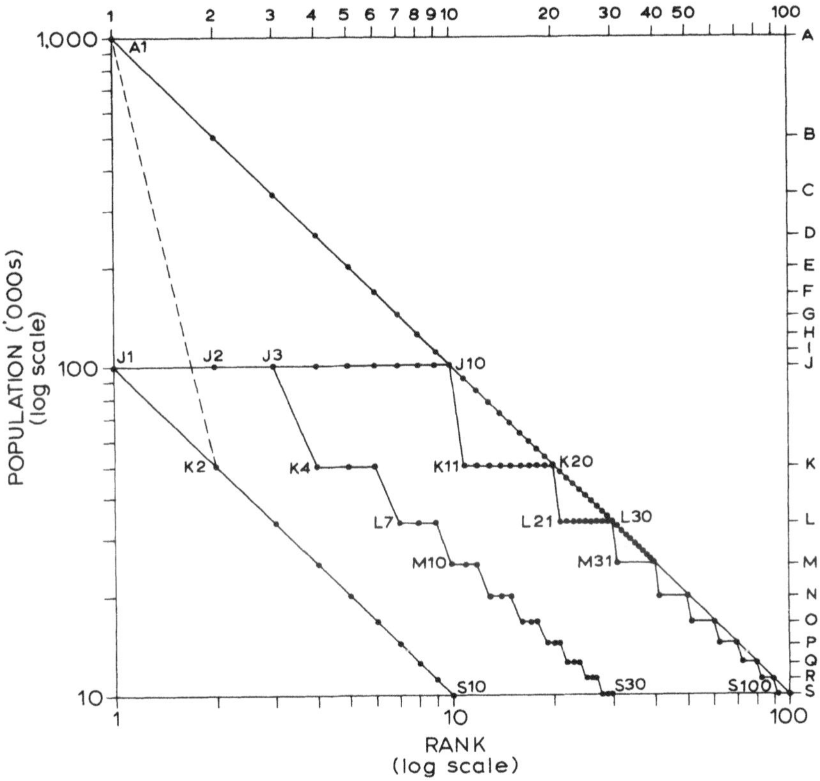

Figure 2.4 An idealized scheme of aggregates of city-systems. *Source:* Harris (1970)

approximately be zero. Only if the ten regions were to be 'fully integrated into a single functionally co-ordinated and interacting urban system' would one expect to find the 100 cities distributed in the rank-size curve of A1, B2 . . . J10 . . . S100. He suggests that a distribution 'with a flat top . . . and several "steps" down the line warns that the cities plotted may represent several relatively independent regions with little functional integration' (Harris 1970, p. 132). Conversely, a distribution in which the curve steepens at the upper end – as in A1, K2, L3 . . . S10 – would suggest that the largest city is disproportionately large because it may 'serve' a much larger area than is encompassed by the cities included within the plot. He uses these ideas to study the size distributions of the Soviet Union and to derive a regionalization of the country based on the shape of the curves of subsets of cities. His partitioning of the country produces twenty-four regions of which no fewer than fourteen have principal cities and boundaries which are identical to those of the major planning regions of the USSR.

Such an exercise is a most interesting attempt to take to its logical conclusion the idea, so widely advocated, that 'lognormality' or rank-size distributions are indicative of complex interacting functional systems. The technique is, however, of little analytical value since it can hardly be hoped to provide a sensitive or diagnostic measure of whether a particular city or set of cities does or does not 'belong' to some hypothesized 'system' or 'sub-system'.[1] Nor, indeed, would it necessarily be the case that the sub-systems would form distinct spatially-discrete areas.

Indeed, in the whole of the work on city sizes, there is much confusion, great difficulty in distinguishing the different distributions, many problems in drawing inferences about processes which might give rise to them and even doubt about the goals of the exercise itself because of the notorious difficulty of arguing about the processes underlying static cross-sectional forms when so often, according to the principle of equifinality, many different processes give rise to identical end-product forms. At no point can one turn to any of the studies of size distributions and claim with confidence that they demonstrate the effectiveness of a systems' approach. As a means of analyzing an array of city sizes, the notion that the set of places comprises a system is an attractive one, but until the objects themselves can be defined *in toto* and until the linkages which determine the relationships between them can be specified, one must accept the conclusion that the existence of regularities in such data suggests relatively little about the substantive phenomena themselves.[2] Kendall (1961, p. 15) puts it rather elegantly:

> It may well be that such things as wars, strikes, epidemics and accidents can all be represented by a common model – a Statistical Theory of Disaster would make an attractive study. It is hard to resist the conclusion . . . that in unifying our observed patterns in this way we are approaching something akin to fundamental laws and moving towards that synthesis which is one of the main goals of scientific endeavour.

But, he adds:

> Nobody would claim we have got that far.

[1] It is interesting that the size distribution of the 100 largest cities in the world as a whole produces exactly the same 'steps' (although not the same slope) as in Harris' idealized diagram. Doubtless only the more heroic would draw conclusions from such an observation.

[2] A conclusion which echoes that of Lotka in 1926. See above, p. 20.

We are forced to a rather pessimistic conclusion about the value of the numerous exercises which have tried to argue that city-size regularities can be taken as evidence of the system-like character of sets of cities. If our concern is with substantive aspects of cities, rather than with probability theory *per se*, the study of size distributions appears to be an elaborate maze which ends only in a cul-de-sac.

CHANGE OVER TIME AND THE PROCESS OF URBAN GROWTH

A brutal, but effective, path out of a maze is to cut through the hedges. Instead of following yet more curves of size distributions, a more profitable avenue, and one which has been surprisingly little explored, is the study of the dynamics of the urban growth which underlies city sizes. Despite the many assumptions about growth rates which have been used, studies of urban growth itself are remarkably few and yet it is the way in which the component cities grow or decline which determines the shape of the size-distribution curves. For empirical data, there is still little detail known about a number of fundamental questions. Do cities in fact grow by a constant but random proportion of their existing sizes? If this 'law of proportionate effect' holds (Gibrat 1931), what are the mechanisms by which it is sustained and what are its implications for our views of urban systems?

The approaches to such questions have mostly adopted rather tangential routes. There has, first, been an attempt to study changes over time in the shape of size-distribution curves. Some of these have suggested that the curves tend to approach towards some form of straight-line logarithmic plot. Bell (1962), for example, suggests this for Israel and Berry (1969) supports the suggestion in his comprehensive review of data from countries in the developing world.[1] Berry concludes that this tendency might support the argument that increasing numbers of a nation's cities are progressively drawn into the ambit of modern urban-orientated market economies and that the suggestion might have planning implications for isolating the constraints within a national economy which obstruct the development of a maximally-probable size distribution and one which betokens the rapid spread of assumedly beneficial growth stimuli throughout the whole hierarchy of places. The study of such changes in the shape of size distributions also gives some rather aggregate indications of the way in which the set of places is growing over time.

A second dynamic approach to urban growth and change has been the study of shifts in the ranks of cities over time. Carter (1969), for example, looks at correlations between the ranks of cities in Wales at successive decades in the nineteenth century. His argument hinges again on a vaguely-specified systems'

[1] Although this tendency for city sizes to approach a straight-line distribution is certainly by no means general. Friedmann (1966a, pp. 146-50), for example, notes the persistence of a 'primate' pattern in Venezuela. This is but one of many counter examples.

approach to the growth of cities since he sees an urban system as evolving through cyclical phases of genesis, transition, sorting and climax in which major technical and organizational changes and exogenous stimuli produce new patterns of regional demand and so lead to further cycling of the genesis sequences. He takes the populations of Welsh towns at each decade from 1801 to 1901, ranks each place from largest to smallest and calculates an overall rank correlation coefficient for each of the intercensal decades. The results demonstrate that the coefficients are markedly lower in the period 1861-81, suggesting that the injection of coal-mining towns into the system in this period corresponds to a new phase of urban genesis which leads eventually to the emergence of a new city system. Such a study of changes in the rank ordering of towns can reveal, albeit tangentially, something of the nature of urban growth and can certainly, as Carter suggests, point in a rather aggregate fashion to periods of marked change in the distribution of individual city growth rates. Other studies which have looked at individual city rank-orders, have pointed to the tendency for shifts in rank to be small at the upper end of the size array, but to be very considerable as one moves to progressively smaller cities. Luckermann (1966) shows that the shifts in rank of the largest cities in the United States conform to this pattern. The same is true of data for British cities. Figure 2.5, for example, plots the ranks of certain cities in England and Wales over the period 1801-1911. The cities which are included are the twenty-five largest places in 1801 and their ranks and populations are shown at each of the twelve successive dates. London, the largest town in 1801, retained its rank throughout the whole period. Manchester and Salford, starting in the second rank, had fallen to third place by the end of the period at the expense of Liverpool which had moved from third to second. Birmingham retained its rank of fourth throughout the whole period. All of these larger places, indeed, show relatively little change in their rank orders over the whole period. On the other hand, with the smaller places there is a tendency for increasingly large fluctuations to occur at smaller city sizes. Some places move rapidly up the rank hierarchy: Leicester, for example, was 21st in 1801, had moved to 13th by 1891 and retained this rank up to 1911. Other places moved very rapidly down the hierarchy: Exeter, for example, began at 20th position, and had fallen to 65th by 1911, even though its fall was halted briefly in the period 1821-41. Other places show more fluctuating fortunes: Coventry, which began in 19th position, had fallen to 48th by 1881, but climbed back to 33rd by 1911.

Analyses of such changes in rank order can reveal much in detail about the demographic fortunes of individual towns and they appear to suggest the ways in which growth occurs in a set of places. The literature of urban growth includes more than a few such instances of the study of changes in rank order (Madden 1956a; Luckermann 1966; Borchert 1967; Lopes 1972; and, a precociously early example, G. G. Chisholm 1897, p. 525). The aggregate patterns which have appeared and which are illustrated by the example of England and Wales, suggest

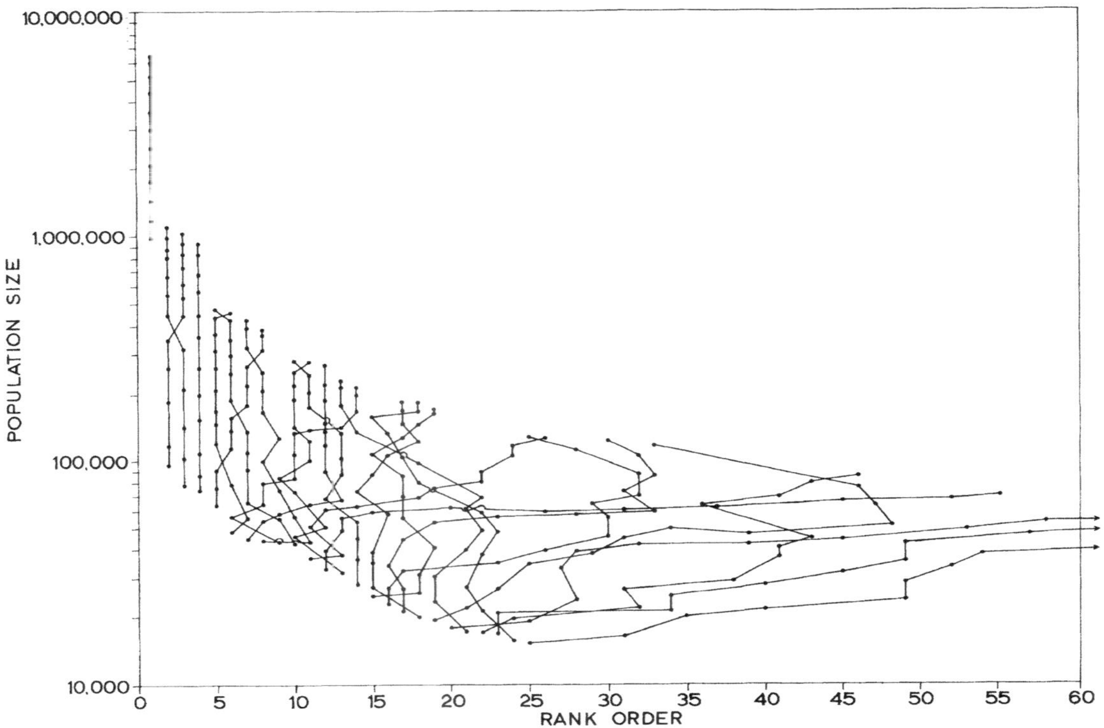

Figure 2.5 Changes in rank order, 1801–1911. Successive ranks of the twenty-five largest towns in England and Wales in 1801. Each line represents one town; the lowest dot shows its rank and size in 1801 and successive dots show its rank and size at each decade up to 1911. The towns (ranked in their 1801 order) are as follows: London; Manchester/Salford; Liverpool; Birmingham; Gloucester; Leeds; Portsmouth; Newcastle/Gateshead; Plymouth; Sheffield; Norwich; Bath; Nottingham; Hull; Sunderland; Medway Towns; Stockport; Bolton; Coventry; Exeter; Leicester; York; Norwich; Oldham; Chester

the stability of the upper end of the size array of places and the great fluctuation of sizes at the lower end. Had fig. 2.5 included more than a mere handful of ranks and had it included places which started at lower ranks and subsequently moved into these higher ranks, this impression would have been reinforced.

This might suggest that a more formal statistical analysis of rank changes over time would be one way in which to study the growth and the fluctuations of growth in a set of cities. The difficulty in such an approach, however, is twofold. First, the degrees of freedom of places to change rank within the hierarchy differ from one rank to another. The largest place of all cannot increase its rank; it can only either retain its first position or fall to a lower rank. If we take a constant set of places, the very smallest place can conversely only retain its rank or increase it; it cannot fall in rank. That place in the very centre of the array would appear to be the only place which has an evenly-balanced probability of plus and minus changes, since there is an equal number of ranks above and below it. Secondly, however, since the size distribution of places will be highly skewed, the absolute difference in population between adjacent ranks at the upper and lower ends of the distribution will be very different. The probability is that there will be a much greater absolute difference in the population sizes of the first- and second-ranked places than between the 100th- and 101st-ranked places; and that the absolute difference in population between the nth-ranked place and the $n - 1$st place will be greater than between the nth and $n + 1$st place. This means that only slight differences in the growth rates of small places might lead to marked changes of rank whereas even marked differences in the growth rates of large places may produce no changes in their ranks. The effect of these two characteristics is that the probable distribution of shifts in the ranks of places will not be normally distributed,[1] and that the probabilities will differ from different initial rank positions. Thus, the probabilities for the largest, smallest and middle places will be highly right-skewed, highly left-skewed and somewhat right-skewed respectively and, while the first two distributions would be highly peaked, the third would be flat topped. This is simply to say that a change of one rank position has very different connotations for a city of 1,000,000 than for a city of 10,000 population.

If an analysis of rank changes were to be attempted, it would therefore be necessary to calculate the different *expectations* of shifts in rank for each different initial rank and these would then form norms against which the *actual* changes in rank could be compared. The task would be complex and probably impossible, but in its absence it would not be possible to interpret the significance of any changes in rank ordering; nor could any pattern of rank shifts, such as the suggestion of stability in the upper end and fluctuation at the lower end of the array of places, be translated with any confidence into suggestions about the variation in growth rates of the set of places.

[1] Thus the centrally-ranked place will *not* in fact have a uniformly distributed probability of moving up or down as was suggested above.

To develop and test ideas about the growth rates of towns, shifts in rank ordering would therefore not appear to be a workable surrogate for a study of the actual population changes of urban areas. We must therefore turn to the actual populations rather than the rank orders of cities to see to what extent Gibrat's law of proportionate effect may hold and what implications this or some other pattern of urban growth may have upon both the theoretical notions of cities and the empirical data of urban sizes. If our interest was primarily in the distribution of city sizes, any findings which were made about the processes of urban growth could then be fed back into the development of appropriate mechanisms by which such distributions were generated and either maintained or altered over time. Enough has been said already to suggest that this is not a topic worth pursuit. Instead, we shall turn to a study of urban growth rates *per se*, ignoring the sterile debate on size distributions – which may form a useful context, but not a profitable goal – and avoiding the wilder claims of the proponents of systems' approaches.

PART 2

Analysis

CHAPTER 3

The growth of cities in England and Wales in the nineteenth century

Facts alone are wanted in life. CHARLES DICKENS

It seems clear that, whatever interest the size distributions of cities may have in themselves, greater interest attaches not to the static curve of sizes, but to the growth processes of towns which first generate and subsequently maintain or alter the curves. The various attempts to derive theoretical distributions have worked from sets of assumptions about these growth processes, but have not explored the extent to which their assumptions may be reasonable or may be met with in the real world. Such an examination seems long overdue, less for the light which it might throw on notions about rank-size curve, than in its own right as telling us something about the nature of urbanization. In this chapter we shall therefore look at data on the growth of towns in England and Wales in the nineteenth century to see to what extent empirical questions about the process of urbanization can be answered as well as to develop some more general ideas about urban growth.

Two preliminary points might be made before turning to an examination of the data. First, if our interest lay in the consequences of urban growth processes on size distributions, by looking at empirical data we would not be able to derive a model of the effect of urban growth on *generating* city sizes; data on the growth of cities will suggest only what some given size distribution may become at a future date.[1] The data could therefore help in understanding the way in which a size distribution might be maintained or might alter over time, but not how it would initially be generated. Second, a word might be said both about the time and the area to be studied. As to the period, it has already been argued that the nineteenth century – Giffen's 'statistical century' – can be regarded as a most useful compromise period in Britain between the pre- and post-industrial periods. Not only does this mean that, so far as our individual units of observation are concerned (call them 'objects' if one will), one can place greater confidence in the necessary assumption that each town is relatively more self-contained, and hence relatively more of a unit or object, than would be the

[1] Unless, of course, one were to demonstrate formally that the model of the growth process would *inevitably* lead to a given size distribution after a sufficiently long time period irrespective of the initial starting distribution – an argument which has been applied to a number of the stochastic growth processes (for example, E. N. Thomas 1967).

case for twentieth-century cities, but it is also the earliest period for which one can have recourse to the demographic data of the censuses. Although the census data are very far from being entirely reliable, they are much more so than any other comparable source and have the inestimable virtue of providing an almost complete coverage of urban places for a long period of time.[1] So far as the area of study is concerned, there are obvious limitations in using only England and Wales as an overall unit. It must be recognized, however, that whatever unit were to be selected, there would always be the problem of the lack of closure of the boundaries of the 'system'. England and Wales were much affected, in terms of their economic growth and the growth of their towns, by concurrent developments in Scotland and in Ireland – more obviously so in the case of the recurrent waves of immigration from the latter unfortunate country. Equally, developments in areas outside the British Isles would have their effects. Nevertheless, in what follows, the two countries will be regarded as forming a closed system, immune from developments from outside. In so doing, considerably less than full justice is done to the complex interlinkages between national economies, but the study of international economic development cannot easily be subsumed in a single parameter within a model of national urban growth. Even within the general framework of economic history there is considerable doubt about the totality of the effect of international linkages. The voluminous literature on 'long waves' in the development of a variety of countries suggests that the shocks to one national economy were echoed elsewhere and the late nineteenth-century indicators of the economies of Britain and North America have led Brinley Thomas (1954) to suggest their linkage within what he calls an 'Atlantic economy' in which the crests and troughs of the two moved in inverse harmony.[2] Saul (1962a), Habakkuk (1962b) and others, however, have doubted the functional connection of the two areas, arguing instead that fluctuations in indices such as house-building are essentially of domestic origin. The case is by no means yet proven and, while there is obviously much room for further study of the effects of international trade upon nineteenth-century England and Wales, it must fall outside the province of this study.

URBAN POPULATIONS

To measure the growth rates of individual towns, one needs to assemble population data for successive times for a number of defined urban areas. This

[1] The various local eighteenth-century surveys, from which much detailed demographic data can be derived, suffer both from the patchiness of their coverage as well as their lack of coincidence in time.

[2] The argument has been extended beyond Britain and America. Schumpeter's (1939) classic work on business cycles drew Germany into the circle and a number of countries of white settlement have also been seen as linked to the fluctuations of Britain from whence their principal funds of capital were derived (see, for example, the discussion in B. Thomas 1972; and, on Australia, the dissenting views of Butlin 1964).

immediately raises a number of the problems of definition which were earlier discussed in connection with what is meant by a 'city'. At what point does a collection of people and buildings become a town; how can one define the areal extent of a given individual town? Such problems have long bedevilled any work which looks at changes in urban populations over time and they account for the uncertainty which must hang over any estimate of the relative sizes of urban and rural populations within any country, no matter how reliable its census information. A tentative definition of a city has already been made which suggested it as an area which contains a relatively large number of people, which is relatively densely settled and in which the majority of the population is involved in activities which are not agricultural. This provides at least guidelines to help resolve the difficulty of classifying urban and non-urban areas. The population data which are used have been taken from the mammoth recalculations of nineteenth-century census returns undertaken by Mr C. M. Law and kindly made available by him. Law (1967) recalculated urban populations for urban areas which were more realistically defined than the administrative areas given in the census returns and, in defining towns for this purpose, he used three criteria which a place needed to meet to qualify as a 'town': a minimum size of population; a minimum density; and a criterion of spatial clustering.

The minimum size is taken at 2,500; a threshold which is large enough to exclude many of the purely mining communities which sprang up in counties such as Durham during the later part of the century and also large enough to exclude most of the small market towns which never attracted industry to them, which relied almost wholly on supporting their tributary agricultural areas and which often suffered radical population decline in the later part of the century. It might be argued that, rather than taking a fixed population size as a criterion of qualifying as a town, a better yardstick might be a variable or sliding size-threshold, so that, as the century progressed, a successively larger figure would be used as a minimum size. This would certainly accord with the realistic argument that the scale of efficient size increases with time as technical and organizational expertise improves and raises the level of scale economies and the lower threshold of efficient operation of units.[1] The use of such a sliding minimum would, however, have two main disadvantages. Not only would it be as difficult to justify the selection of any particular threshold at a given time as to select a constant threshold, but it would also introduce difficulties in handling towns whose population, even though remaining static or growing only slowly, would yet 'die' out of the urban 'system' as the threshold became progressively higher. As Friedlander comments with reference to just this problem of varying minimum size thresholds, 'It seems that there are no simple or unique answers to these questions and there could probably be good arguments in favour of different approaches to such problems' (Friedlander 1970, p. 423).

[1] A theme which forms the underpinnings of much of the argument of Social Area Analysis which has looked at changes in the internal structure of both cities and national economies (see, for example, Shevky and Bell 1955; Udry 1964; McElrath 1968).

Law's density criterion was fixed at a figure of one person per acre which, as he says, would be a very low figure for urban areas, but a very high figure for rural areas. It is thus a useful yardstick for deciding which of the parishes or areas surrounding an administrative urban area should be included in the total urban population or, indeed which administratively-defined non-urban areas should be regarded as towns. Since the censuses for England and Wales give population data for administrative units which are often seriously underbounded as descriptions of urban areas, it was this criterion which was used to define the basic areal extent of urban areas.[1]

Finally, the nucleation criterion takes account of the attempt to define towns as spatially continuous built-up areas. In a number of the parishes of England and Wales, even though the criteria of minimum population size and minimum density are met, the population is dispersed in a series of small communities over a relatively wide area and the several communities are not physically contiguous. In the nineteenth century, this was particularly the case of many settlements in South Wales and parts of northern England, where, most usually, mining communities were found scattered discretely throughout large parishes. In such cases these populations would not be regarded as a single urban area even though meeting the first two minima.

THE DATA

These three criteria therefore underlie the redefinition of urban places used throughout this study, and the population figures on which the subsequent analyses are based depend critically upon the validity of the criteria. They depend as well upon the varying accuracy of the census returns. Undoubtedly, the census improved over time. It is generally agreed that the first two censuses, in 1801 and 1811, failed to include a significant proportion of the population although estimates of that proportion differ (Taylor 1951; Krause 1958). Thereafter, the census figures are much more reliable, but by no means perfect. A marked improvement came after 1841 when a new approach to the collection of the data was instituted. In the first three censuses, information was provided by Overseers of the Poor in the several parishes – a somewhat casual exercise which suffered from a number of potential dangers, not the least of which was

[1] In the United States, the Bureau of the Census – for long much more alive to the inadequacy of purely administrative definitions of urban areas than the British Census Office – began to experiment with more extensive definitions of urban aggregates as early as 1910 when statistics were given for 'Metropolitan Districts'. In 1950, data were reported both for 'Urbanized Areas' and 'Standard Metropolitan Areas'. Subsequently, the 'Standard Metropolitan Statistical Area' was added (for definitions, see U.S. Bureau of the Budget 1967 and, for the 'Urbanized Area', Carter 1972, p. 24). The nearest British equivalent – the 'conurbation' – falls between these two American definitions and suffers from the fact that it is still composed of administrative units.

that the whole operation might be spread over a number of days thus increasing the risk of double counting. The *Report* on the 1841 census (*Parlimentary Papers*, 1843, XXII, 5), in evaluating the role of the Overseers, considered them to have been 'a body of men, in many parts of the country, fully competent to the task thus imposed upon them, and anxious to discharge it to the best of their ability; but in some districts, from want of education, or habits of business, and from ignorance of the importance and real object of the inquiry they were conducting, essentially unfitted for it.' For the 1841 census the country was divided by local Registrars into as many as 35,000 small Enumeration Districts, each of which was assigned to one enumerator who was required to undertake the enumeration 'in one day'. The 1841 census also postdated the reform of urban areas effected by the Municipal Corporations Act which meant that many new urban areas were recognized as administrative units for the first time and the boundaries of all the boroughs and municipalities were faithfully demarcated in maps signed in the spidery hand of R. K. Dawson (*Parliamentary Papers*, 1837, XXV-XXVIII). From 1851 onwards, the census authorities also began to produce tabulated statistics of urban and rural populations helping at least to sharpen the awareness of the distinction between what were then considered as urban and non-urban areas.[1]

The two most troublesome of the urban definitions are, of course, those of the very largest and the smallest towns. As the nineteenth century progressed, the largest towns grew to incorporate once-separate peripheral communities. Given the definitions adopted, had the peripheral boroughs or communities been regarded as discrete places before being submerged into the sprawl of their neighbouring giants, once they *had* been submerged they would have 'died' as towns and the larger places correspondingly would have appeared to increase rather suddenly during the decade in which the amalgamation was assumed to have occurred. The largest urban areas have therefore often been amalgamated with certain of their neighbouring communities to produce a generous definition and the population totals of the adjacent areas have been included in the overall total once the density criterion outlined above had been met. Thus, the definition of London traces out an area which, at its most considerable extent, includes, in addition to the Administrative County, a number of peripheral areas which can be considered to have become part of the metropolis by 1911 – the end of the period of study. In the north for example, Southgate, Edmonton, Tottenham, Wood Green, Hornsey, Finchley and Hendon came to be included; in the east, Woodford, Walthamstow, Wanstead, Leyton, Ilford, Dagenham, Barking, and East and West Ham; in the south, Bromley, Penge, Beckenham, Croydon, Mitcham, Merton and Wimbledon; and in the west, Richmond, Barnes,

1 Accounts of the successive census changes are given in Interdepartmental Committee on Social and Economic Research 1951, and, especially, in Drake 1972.

Mortlake, Chiswick, Brentford and Ealing, Acton, Hanwell and Willesden.[1] Certain of the large provincial northern towns also include peripheral boroughs or communities. For example; Manchester eventually grew to include not only Salford, but also Stretford, Droylesdon, Prestwich and Failsworth; Liverpool includes Birkenhead; Sheffield includes Handsworth; Birmingham includes Smethwick; and Newcastle includes Gateshead.

For the smaller places, the main difficulty lies in estimating the early populations of certain places which grew very suddenly. With the very rapid urban growth of the nineteenth century – a growth for which the experience and the administrative machinery of early Victorian England were largely unequipped to deal – a number of difficulties occur in the calculation of the populations of towns which are 'born' into the system when they reach a figure of 2,500 or over. The great majority of places which were not already 'towns' by 1801, were 'born' with populations of less than 5,000 and the progress of their subsequent rises and falls in population can be taken with some confidence. A number of places, however, first appear with populations considerably larger than this and it is clear that for some of these the inadequacies of the census data mean that it is impossible to trace the early history of their demographic expansion. This means that there will be some undercounting of the real number of smaller 'towns' and that the early growth rates of such places cannot be included in the subsequent calculations. Table 3.1 lists all those places which were 'born' with populations of 10,000 and over and also a regional distribution of the total numbers of places 'born' at 5,000 or over. It is clear that the great majority of these places is comprised of two types of town. First are the mining towns of Durham, Northumberland and South Wales and the manufacturing towns of Lancashire, the West Riding and the Black Country. Many of these places grew very rapidly from very small origins and some tended to be designated as separate administrative areas only at a relatively late stage in their growth so that it is impossible to attach population figures to them until a sufficiently fine administrative net was created. Second are the towns which became dormitory suburban areas very late in the nineteenth century or in the first decades of the twentieth. This accounts for the relatively large numbers of places 'born' at suspiciously high figures in the counties surrounding London.

Not all of these places are reflections of the inadequacies of the census information, however. A number of Victorian towns were genuinely 'born' with

[1] In the light of this broad definition of London, it is interesting to note Hobsbawm's discussion of London as a labour market. Unlike other conurbations, London developed a single wage rate which applied theoretically to the whole urbanized area throughout the nineteenth century. Its extent, understandably, widened throughout the period: for example, in 1834, the tailors' union claimed jurisdiction over an area with a four-mile radius from Charing Cross; in 1865, the compositors claimed an area with a radius of fifteen miles. This was markedly different from provincial conurbations. South-east Lancashire, for example, was still a mosaic of trade-union districts with different wage rates as late as 1906. Tyneside, in 1806, was divided into three districts of the Amalgamated Society of Engineers and five of the relevant semi-skilled union (Hobsbawm 1964, pp. 4-5).

Table 3.1 *Towns entering the urban 'system' with large initial populations*

A. Towns 'born' with populations of 10,000 or over

Town	*Reference*	*Date 'born'*
Staleybridge	CHE 29	1851
Glossop	DER 17	1851
Jarrow	DUR 24	1871
Shotton	DUR 36	1911
Aldershot	HAM 1	1861
Bournemouth	HAM 5	1871
Bacup	LAN 8	1861
Eccles	LAN 27	1871
Heywood	LAN 35	1831
Mossley	LAN 54	1871
Coalville	LEC 3	1891
Enfield	MDX 2	1861
Oakengates	SAL 7	1891
Burslem	STA 6	1831
Coseley	STA 9	1871
Sedgley	STA 21	1871
Chertsey	SRY 6	1891
Saddleworth	YWR 88	1871
Stanley	YWR 103	1891
Abercarn	MON 1	1891
Abersychan	MON 3	1871
Ebbw Vale	MON 8	1871
Aberdare	SWA 9	1851
Barry	SWA 10	1891
Caerphilly	SWA 14	1901
Gelligear	SWA 15	1901
Ogmore	SWA 20	1901
Pontypridd	SWA 22	1871
Rhondda	SWA 25	1871

B. Distribution of all towns 'born' with populations of 5,000 or over

(i) Distribution by date *(ii) Distribution by county*

Date	*No. 'born'*	*Area*	*No. 'born'*	*Area*	*No. 'born'*
1801	0	Buckingham	1	Northumberland	3
1811	0	Cambridge	1	Salop	1
1821	0	Cheshire	4	Stafford	6
1831	5	Derby	5	Surrey	7
1841	8	Durham	14	Sussex	2
1851	10	Essex	1	Warwick	2
1861	9	Hampshire	3	Worcester	2
1871	36	Hertford	2	Yorks (NR)	4
1881	6	Kent	2	Yorks (WR)	9
1891	18	Lancashire	10	Monmouth	10
1901	6	Leicester	1	N. Wales	2
1911	8	Lincoln	1	S. Wales	10
		Middlesex	3		
Total	106			Total	106

very high populations. Middlesbrough is perhaps the classic example of a newly-created Victorian town. It enters the list of 'towns' – and hence of the calculations of this study – in 1841 at a population of almost 6,000 and by 1861 had reached nearly 20,000. Three further examples can be quoted as illustrations of the mining, industrial and residential categories which comprise the great bulk of the genuinely mushrooming towns which form part of table 3.1. Ashington, in Northumberland, shared in the late nineteenth-century growth of parts of the concealed coalfield of the North East and its population figures read as follows: 1871 – 1,002; 1881 – 2,091; 1891 – 5,307; 1901 – 13,972; and by 1911 it had almost reached 25,000. It was therefore 'born' in 1891 at the very high figure of over 5,000. Scunthorpe grew on the basis of its iron workings in the last quarter of the nineteenth century and its population figures read: 1871 – 1,852; 1881 – 5,173; 1891 – 7,255; and by 1911 it had reached almost 20,000. Royal Leamington Spa developed as a resort and dormitory area much earlier in the century and its figures read: 1821 – 2,183; 1831 – 6,209; 1841 – 12,864; and by 1911, after a period of stability and even slight decline, it stood at over 26,000. In each of these three cases the particular town – illustrating a different and typical type of mushrooming growth – entered the system at figures well above the critical threshold of 2,500, but the first population figure can be used with some confidence as the earliest datum for that place as a 'town'. In other cases where towns enter at a high population figure, the first population datum must be treated with less confidence and the early figures are simply impossible to derive because of the inappropriateness and inflexibility of the administrative boundaries for which data are reported in the census. This would be the case, for example, with Bournemouth which is treated as a single aggregate together with Poole. The joint population figures can only be derived from 1881 onwards. Before that date, even though separate figures for Bournemouth alone could have been used, the subsequent addition of Poole would have created a quite artificial boosting of population at 1881, so that the aggregate 'town' is considered as having been 'born' at a figure of just over 30,000 in 1881.

Problems of definition of this sort are an inevitable conjunct of any work on urban growth. Despite such difficulties, and despite the variable quality of the census sources from which the populations have been recalculated, Law's figures provide as sensitive and accurate a set of estimates of urban populations as one could hope to derive.[1] Certainly in that they apply to areas which approximate as closely as possible to the actual and changing extent of the built-up area of nineteenth-century towns, they provide better estimates of urban populations than do the raw census data on which so many previous studies have had to be

[1] Only an impossibly laborious calculation based on the information contained in enumerators' books could provide a more sensitive set of populations and this source is, of course, available only for census dates between 1841 and 1871.

Table 3.2 *Numbers of towns in different size groups*

Date	Size Group ('000s)																		
	2·5–4·9		5–9·9		10–19·9		20–39·9		40–79·9		80–159·9		160–319·9		320–639·9		⩾ 640		Total
	No.	%	No.	%	No.	%	No.	%	No.	%	No.	%	No.	%	No.	%	No.	%	No.
1801	160	62·50	47	18·36	32	12·50	7	2·73	8	3·13	1	0·39	0	–	0	–	1	0·39	256
1811	190	61·69	64	20·78	31	10·06	12	3·90	7	2·27	3	0·97	0	–	0	–	1	0·32	308
1821	200	55·40	89	24·65	38	10·53	20	5·54	9	2·49	3	0·83	1	0·28	0	–	1	0·28	361
1831	207	50·24	115	27·91	44	10·68	26	6·31	13	3·16	4	0·97	2	0·49	0	–	1	0·24	412
1841	226	48·60	121	26·02	62	13·33	32	6·88	13	2·80	7	1·51	2	0·43	1	0·22	1	0·22	465
1851	234	44·91	135	25·91	85	16·31	31	5·95	23	4·41	8	1·54	2	0·38	2	0·38	1	0·19	521
1861	255	43·97	153	26·38	91	15·69	40	6·90	24	4·14	9	1·55	4	0·69	3	0·52	1	0·17	580
1871	274	39·65	201	29·09	111	16·06	55	7·96	28	4·05	13	1·88	5	0·72	2	0·29	2	0·29	691
1881	270	36·00	215	28·67	135	18·00	65	8·67	37	4·93	17	2·27	7	0·93	1	0·13	3	0·40	750
1891	293	35·13	230	27·58	159	19·06	77	9·23	41	4·92	20	2·40	8	0·96	2	0·24	4	0·48	834
1901	290	32·77	225	25·42	190	21·47	89	10·06	49	5·54	23	2·60	11	1·24	4	0·45	4	0·45	885
1911	274	29·69	243	26·33	193	20·91	109	11·81	56	6·06	29	3·14	11	1·19	4	0·43	4	0·43	923

Note: Percentage figures are percentages of the total number of towns at each date.

based – as, for example, in the figures given by Mitchell (1962, pp. 24-7) which have been widely quoted elsewhere (for example, Banks 1968).[1]

THE NUMBERS OF URBAN PLACES

If we look by way of introduction at the set of places which these definitions produce during the period 1801-1911, the easiest way to familiarize ourselves with the material is simply to look at the changing numbers of towns in each

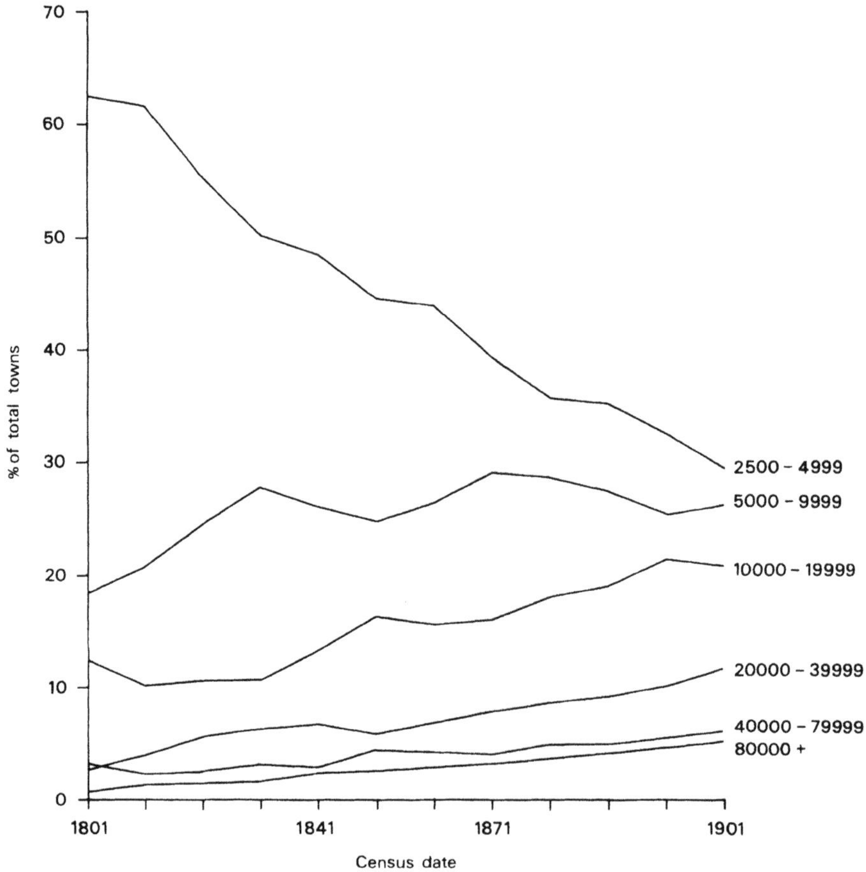

Figure 3.1 Percentages of towns in different size groups in England and Wales, 1801–1911. Details are given in table 3.2

[1] The same would be true of the heroic calculations of nineteenth-century commentators such as A. F. Weber (1899) and Welton (1900). In calculating percentages of the urban population for England and Wales as a whole, Law (1967) compares the results based on his urban definition with the figures suggested by the census itself and by Price-Williams (1880), Welton (1900), A. F. Weber (1899), and Vince (1955).

population size class over the whole span of the twelve dates. Table 3.2 shows the number of towns in each of nine size classes which have been selected so that each size class is a constant factor of twice that of the previous class. It can be seen first of all that the number of towns increases rapidly from a mere 256 in 1801 to as many as 923 in 1911. Secondly, within this increasing number of places, the proportional representation of each size class changes in a fairly systematic fashion. The percentage of towns in the size range 2,500-4,999 falls progressively from over 60 to under 30; that in the size range 5,000-9,999 rises from less than 20 to almost 30 and then falls to level off at about 25; percentages in the size ranges above 10,000 tend almost universally to increase throughout the whole span of the period. Figure 3.1 traces these changes, showing the relatively greater rise in the numbers of large places at the expense of the smaller.

CHANGES IN SIZE

These are simple descriptive statistics which give some preliminary idea about the numbers of places which are to be studied and also suggest that there may be regularities in the changing sizes of those places which could lead to the development of hypotheses about urban growth. It is not, of course, the number of towns in which our interest lies, but the population changes of individual size groups of towns at successive decadal intervals. The first, and simplest, way in which this might be studied is to plot the movements of individual towns between the eight size classes during each pair of census dates. These are shown in table 3.3 which gives transition matrices for each of the eleven successive periods. Each matrix shows the number of places which start the decade with populations in one size class and end the decade with populations either in the same or some different size class. The row entries show, for each size class, the classes in which towns ended the decade; the column entries show, for each size class at the end of the decade, in which size class the places had started the decade. In the first decade, for example, 160 places had populations in the size class 2,500-4,999 in 1801; by 1811, of these 160 places only 135 were still in that size group, twenty-four had increased their populations into the size group 5,000-9,999, one place had lost population so as no longer to qualify as a 'town', and so forth. The totals of the rows and columns can be interpreted in the same way: the totals of the rows show the number of places in each size group in 1801; the totals of the columns show the numbers in each size group in 1811.

Attention might be drawn to four points in these matrices. First, the range of movement between categories is relatively small. This is very much a function of the large ranges of population which are included in each size category. For a place to move from the mid-point of one size group to the mid-point of the next higher group, it would need to double its population during the course of the decade. It is therefore not surprising that there is only a minority of cases in

Table 3.3 *Transition matrices: numbers of towns in population size groups at successive dates*

		No. in 1811									
Size group		*0*	*1*	*2*	*3*	*4*	*5*	*6*	*7*	*8*	*Total*
	0	625	53	0	0	0	0	0	0	0	
No.	*1*	1	135	24	0	0	0	0	0	0	160
in	*2*	0	2	40	5	0	0	0	0	0	47
1801	*3*	0	0	0	26	6	0	0	0	0	32
	4	0	0	0	0	6	1	0	0	0	7
	5	0	0	0	0	0	6	2	0	0	8
	6	0	0	0	0	0	0	1	0	0	1
	7	0	0	0	0	0	0	0	0	0	0
	8	0	0	0	0	0	0	0	0	1	1
Total			190	64	31	12	7	3	0	1	308/256

		No. in 1821									
Size group		*0*	*1*	*2*	*3*	*4*	*5*	*6*	*7*	*8*	*Total*
	0	573	53	0	0	0	0	0	0	0	
No.	*1*	0	147	42	1	0	0	0	0	0	190
in	*2*	0	0	47	17	0	0	0	0	0	64
1811	*3*	0	0	0	20	11	0	0	0	0	31
	4	0	0	0	0	9	3	0	0	0	12
	5	0	0	0	0	0	6	1	0	0	7
	6	0	0	0	0	0	0	2	1	0	3
	7	0	0	0	0	0	0	0	0	0	0
	8	0	0	0	0	0	0	0	0	1	1
Total			200	89	38	20	9	3	1	1	361/308

		No. in 1831									
Size group		*0*	*1*	*2*	*3*	*4*	*5*	*6*	*7*	*8*	*Total*
	0	522	46	3	2	0	0	0	0	0	
No.	*1*	0	161	39	0	0	0	0	0	0	200
in	*2*	0	0	72	17	0	0	0	0	0	89
1821	*3*	0	0	1	25	12	0	0	0	0	38
	4	0	0	0	0	14	6	0	0	0	20
	5	0	0	0	0	0	7	2	0	0	9
	6	0	0	0	0	0	0	2	1	0	3
	7	0	0	0	0	0	0	0	1	0	1
	8	0	0	0	0	0	0	0	0	1	1
Total			207	115	44	26	13	4	2	1	412/361

Table 3.3–*Continued*

	Size group				No. in 1841						
		0	1	2	3	4	5	6	7	8	Total
	0	468	46	8	0	0	0	0	0	0	
No.	1	1	175	31	0	0	0	0	0	0	207
in	2	0	5	82	27	1	0	0	0	0	115
1831	3	0	0	0	34	10	0	0	0	0	44
	4	0	0	0	1	21	4	0	0	0	26
	5	0	0	0	0	0	9	4	0	0	13
	6	0	0	0	0	0	0	3	1	0	4
	7	0	0	0	0	0	0	0	1	1	2
	8	0	0	0	0	0	0	0	0	1	1
Total			226	121	62	32	13	7	2	2	465/412

	Size group				No. in 1851						
		0	1	2	3	4	5	6	7	8	Total
	0	413	46	7	2	1	0	0	0	0	
No.	1	0	187	39	0	0	0	0	0	0	226
in	2	0	1	89	31	0	0	0	0	0	121
1841	3	0	0	0	52	10	0	0	0	0	62
	4	0	0	0	0	20	12	0	0	0	32
	5	0	0	0	0	0	11	2	0	0	13
	6	0	0	0	0	0	0	6	1	0	7
	7	0	0	0	0	0	0	0	1	1	2
	8	0	0	0	0	0	0	0	0	2	2
Total			234	135	85	31	23	8	2	3	521/465

	Size group				No. in 1861						
		0	1	2	3	4	5	6	7	8	Total
	0	350	54	6	3	0	0	0	0	0	
No.	1	4	196	34	0	0	0	0	0	0	234
in	2	0	5	110	20	0	0	0	0	0	135
1851	3	0	0	3	68	14	0	0	0	0	85
	4	0	0	0	0	26	5	0	0	0	31
	5	0	0	0	0	0	19	4	0	0	23
	6	0	0	0	0	0	0	5	3	0	8
	7	0	0	0	0	0	0	0	1	1	2
	8	0	0	0	0	0	0	0	0	3	3
Total			255	153	91	40	24	9	4	4	580/521

Table 3.3–*Continued*

Size group						*No. in 1871*					
		0	*1*	*2*	*3*	*4*	*5*	*6*	*7*	*8*	*Total*
	0	241	77	25	9	2	0	0	0	0	
No.	*1*	2	197	52	4	0	0	0	0	0	255
in	*2*	0	0	124	28	1	0	0	0	0	153
1861	*3*	0	0	0	70	20	1	0	0	0	91
	4	0	0	0	0	32	8	0	0	0	40
	5	0	0	0	0	0	19	5	0	0	24
	6	0	0	0	0	0	0	8	1	0	9
	7	0	0	0	0	0	0	0	4	0	4
	8	0	0	0	0	0	0	0	0	4	4
Total			274	201	111	55	28	13	5	4	691/580

Size group						*No. in 1881*					
		0	*1*	*2*	*3*	*4*	*5*	*6*	*7*	*8*	*Total*
	0	184	53	6	0	0	0	0	0	0	
No.	*1*	0	214	58	2	0	0	0	0	0	274
in	*2*	0	3	149	49	0	0	0	0	0	201
1871	*3*	0	0	2	84	24	1	0	0	0	111
	4	0	0	0	0	41	14	0	0	0	55
	5	0	0	0	0	0	22	6	0	0	28
	6	0	0	0	0	0	0	11	2	0	13
	7	0	0	0	0	0	0	0	5	0	5
	8	0	0	0	0	0	0	0	0	4	4
Total			270	215	135	65	37	17	7	4	750/691

Size group						*No. in 1891*					
		0	*1*	*2*	*3*	*4*	*5*	*6*	*7*	*8*	*Total*
	0	98	68	12	6	0	0	0	0	0	
No.	*1*	2	221	46	1	0	0	0	0	0	270
in	*2*	0	4	168	43	0	0	0	0	0	215
1881	*3*	0	0	4	108	23	0	0	0	0	135
	4	0	0	0	1	54	10	0	0	0	65
	5	0	0	0	0	0	31	6	0	0	37
	6	0	0	0	0	0	0	14	3	0	17
	7	0	0	0	0	0	0	0	5	2	7
	8	0	0	0	0	0	0	0	0	4	4
Total			293	230	159	77	41	20	8	6	834/750

Table 3.3–*Continued*

Size group		*No. in 1901* 0	1	2	3	4	5	6	7	8	*Total*
	0	41	53	3	3	0	0	0	0	0	
No.	*1*	8	231	53	1	0	0	0	0	0	293
in	*2*	0	6	168	56	0	0	0	0	0	230
1891	*3*	0	0	1	129	29	0	0	0	0	159
	4	0	0	0	1	60	16	0	0	0	77
	5	0	0	0	0	0	33	8	0	0	41
	6	0	0	0	0	0	0	15	5	0	20
	7	0	0	0	0	0	0	0	6	2	8
	8	0	0	0	0	0	0	0	0	6	6
Total			290	225	190	89	49	23	11	8	885/834

Size group		*No. in 1911* 0	1	2	3	4	5	6	7	8	*Total*
	0	7	34	7	1	0	0	0	0	0	
No.	*1*	4	238	43	5	0	0	0	0	0	290
in	*2*	0	2	191	31	1	0	0	0	0	225
1901	*3*	0	0	2	156	32	0	0	0	0	190
	4	0	0	0	0	76	13	0	0	0	89
	5	0	0	0	0	0	43	6	0	0	49
	6	0	0	0	0	0	0	23	0	0	23
	7	0	0	0	0	0	0	0	11	0	11
	8	0	0	0	0	0	0	0	0	8	8
Total			274	243	193	109	56	29	11	8	923/885

Note: Size groups are as follows:

0	< 2,500	5	40,000–79,999
1	2,500–4,999	6	80,000–159,999
2	5,000–9,999	7	160,000–319,999
3	10,000–19,999	8	⩾ 320,000
4	20,000–39,999		

which towns appear in cells more than one away from the diagonal. On the whole, such large movements tend to be restricted to the smallest range of town sizes which suggests that it is only in these towns that one might expect to find very high percentage growth rates. Second, it can be seen that the matrices are all 'absorbing' matrices, that is to say that it is not possible for a town to move from one size class to every other of the size classes within the matrix – a requirement which would have to be met for the matrices to be 'regular Markov matrices' (Kemeney and Snell 1960). For this to be possible, the rows would need to have entries in the cells both above and below the diagonal. Instead – and particularly in the higher population sizes – the pattern is for towns either to continue in the same size class or to move up into the next higher size class.

Thus, if a given initial distribution of town sizes were to be powered up by successive multiplications by the transition probabilities,[1] all of the towns would eventually end up in the highest size class since, once above a certain population size, the probability of moving *down* into a lower size group becomes remote or non-existent. Third, the pattern of town 'births' is of interest. These 'births' are shown in the entries given in the 'zero' row and which fall in columns other than the 'zero' column – in other words, places which start a decade at populations below 2,500, but which end with populations greater than that figure. The rapid growth in the number of such new towns has already been noted, but here the pattern is shown much more clearly. The great majority is 'born' in the lowest size category with populations between 2,500 and 4,999, but many are 'born' in the later decades at figures well above this and it is some of these places whose early population figures are impossible to derive (see above, pp. 50–2). Fourth, and conversely, the entries in the 'zero' columns which do not fall in the 'zero' rows show the pattern of 'deaths' of towns – those places which fall below the threshold of 2,500. In most cases these represent places whose population hovered uncertainly around the threshold figure and declined below it for some of the twelve dates. Lyme Regis, in Dorset, for example, grew to over 2,500 by 1831 and reached a peak of 2,852 in 1851, but then declined so as to fall below the threshold in 1881 and only reappeared above it in 1911. Similarly, Dursley in Gloucestershire had a population above 2,500 from 1811 to 1851, but then fell below and again reappeared in 1911. Market Rasen and Alford, in Lincolnshire, both grew above the threshold only by 1861 and then fell below it in 1881 and 1891 respectively and had not reappeared again by 1911. The overall number of such places which 'died' – or fell out of the list of 'towns' – is relatively small. They present far fewer operational problems in deriving population totals than do some of the places which were 'born', not only because of the smaller number involved, but also because, when an administrative core had once been defined for such places, appropriate figures tend subsequently to be given in the census returns.

These are the main features of the transition matrices. In general they show that the great bulk of towns cluster along the diagonals of the tables, confirming in a very approximate way an expectation that urban growth would tend to occur at a rate common to the whole set of towns. A sharper focus is needed, however, before anything more precise or of greater interest than this general description might emerge. There are two possible avenues which could be explored: first the formal use of Markov methods in analyzing the structure of these matrices; and second the adoption of a regression format in looking at the relationship of opening and closing sizes at successive dates.

[1] The transition probabilities would be given simply as a weighting of the raw observations of the matrices. Thus, for each row, the probabilities would be the cell entries divided by the row total. These would give the probability of staying in the same size class or moving to another by the end of the decade.

MARKOV METHODS

The use of Markov methods was not pursued, even though there are some interesting examples of their use in closely similar fields. Markov methods involve the use of transition matrices of change as probabilities by which an initial distribution of observations is successively transformed over a number of iterations until the distribution converges on a terminal distribution vector which represents an equilibrium position to which the probabilities tend irrespective of the values of the initial vector. Alternatively, the number of iterations which are needed to reach the final absorbing state – when all the observations will have fallen into a group from which they do not again emerge – can be calculated for an absorbing matrix. Fuguitt (1965) has used such approaches in looking at population changes in small towns in the United States. For decades in which the pattern of size changes produces regular stochastic matrices, the equilibrium vectors produced by successive applications of the transition probabilities are derived and comparisons between them suggest that markedly different transition probabilities apply from one decade to another. The same differences are also seen by comparing the length of time taken to reach absorption: of the eight decades studied, the decade 1890-1900 shows the most rapid change, reaching absorption in the equivalent of 350 years and with 130 years being the average time which it would take for a place to move from a population of less than 500 to over 5,000; by contrast, the decade 1920-30 showed the slowest change with figures of 4,510 and 3,820 years respectively. In Britain, Lever (1972) has also used Markov methods to examine urban growth for the period 1951-71. He tries to isolate 'most favoured' town sizes by calculating the final vectors of town sizes which result by successive applications of the transition probabilities of 1951-61 and 1961-71. Again, like Fuguitt, he discovers that the transition probabilities differ between the two decades[1] and, in any case, his derivation of a terminal vector of sizes is made less convincing by the fact that his initial matrices contain very few observations and that the matrices themselves consist of a number of absorbing sub-matrices rather than single regular matrices.

Markov methods have been widely used in studies comparable to these of the changes of town size. In the literature of the social sciences, the phenomenon of social mobility has been studied in this way (Prais 1955; Beshers and Laumann 1967), but the closest parallel is in the study of business-firm sizes, the most comprehensive work being that of Adelman (1958). Its particular interest here lies in the way in which Adelman deals with the troublesome categories of observations which are 'born' and which 'die'. Most of the applications of

[1] The fact that marked differences occur between the probabilities of successive decades means, of course, that the successive application of one set of probabilities over a long period of time must be seen in terms of its analytic rather than its predictive value.

Markov methods have ignored these categories and deal instead with an unchanging number of observations – be they towns, firms or whatever. Adelman, however, incorporates them by including in the transition matrices a very large 'reservoir' from which new firms can be created and into which existing firms can disappear. This would be the equivalent, in our transition matrices, of the 'zero-zero' cell in which appear all the places which are less than 2,500, but which are larger than that at some date in the period 1801-1911. Adelman demonstrates that the size of the entry in this cell makes no difference to the eventual equilibrium vector which is produced by the transition probabilities.[1]

Despite the interest of such approaches, Markov methods were ignored here partly because of the unreasonableness of using a simple assumption that the transition probabilities would remain constant over a long period of time, but more particularly because, as has already been noted, all of the matrices were absorbing and thus it would have been possible to calculate only the expected time to absorption for the probabilities of each decade. This, by itself, would have produced somewhat of a sledgehammer statistic which would tell one little about the structure of population changes besides the speed of the overall rate of change.

REGRESSION OF OPENING AND CLOSING SIZES FOR EACH DECADE

Instead, a regression approach would seem of much greater value as a first exercise in looking at the details of the nineteenth-century growth processes which are hinted at in the transition matrices. It was suggested in the last chapter that one might hypothesize that towns grow by some common overall rate, but that the common rate is subject to fluctuations from one town to another. This hypothesis suggests a regression formulation of the following sort:

$$X_t = a + bX_{t-1} + \epsilon; \qquad [3.1]$$

where X_t is the population of a town at time t; a and b are constants; and ϵ is a normally-distributed error or disturbance term. If the growth of the set of places were to conform to the law of proportionate growth, the value of b would be expected to be 1·0 and the overall change produced by this growth within the total population of towns would be represented by the constant a which would measure the vertical displacement of the curve. The variances in the growth rates of individual towns would be contained in the error term. The interest in this formulation is twofold: first in the extent to which real-world data do produce a value of b which is not significantly different from 1·0; and second in the increasing concentration produced in the distribution of urban populations by the value of both the b term and the disturbance term.

[1] The difficulty of dealing with a variable rate of 'births' and 'deaths' is discussed below, pp. 69–70.

CONCENTRATION OF URBAN POPULATION

Before turning to the statistical regression, we might digress a little to expand on this latter point about the growing concentration of population. If it were to be found that empirical data on urban growth fitted a regression model in which the value of b was greater than 1·0, this would mean that larger places tended to grow more rapidly than smaller places and the consequence on the size distribution of those places would be that the slope of the frequency curve would grow steeper as increasing proportions of the total population were contained within the larger cities at the expense of the smaller. There is, indeed, some evidence that the slopes of more developed countries tend in general to be somewhat steeper than those of less developed countries. Lopes (1972), for example, in criticizing the distinction between 'lognormal' and 'primate' size distributions, suggests that attention should instead be focused on the relationship between the degree of economic development of countries and the *slope* of their size-distribution curves.[1] He finds that the correlations between the slope coefficients and indices of economic development for a number of countries are positive and quite high and, while the calculation of slopes is very much dependent upon the lower threshold of size at which cities are selected, the results would tend to support the notion of increasing urban concentration over time. In practice, however, the great majority of studies of rank-size distribution curves has found that the curves at successive dates tend to be parallel with each other and tend to settle down at a slope of approximately 45°.[2] In the regression format of the closing size on the opening size of cities, if we find a b value of 1·0, will this then explain the parallelism of size curves?

In fact, so long as there is a disturbance term around the common growth rate, even a value of b equal to 1·0 will still produce increasing population concentration over time. The connection of concentration and the nature of urban growth is best seen with a simple example. Suppose one takes a set of 128 cities each of which has an identical population at some given time t_1 – a situation perhaps representing an early stage of unspecialized development in which each place is relatively independent and serves only a small and relatively self-contained tributary area (a statistical parody, in fact, of the pre-industrial city). Let us assume that each city has a population of 10,000; then a measure of concentration which might be used is the percentage of the total population contained in the ten largest places and this, at time t_1, would be 7·8. Obviously if each individual city were subsequently to grow by an identical rate, the percentage of population contained in the ten largest places would remain unchanged no matter how many time periods were involved. However, if we

1 A suggestion which was advanced much earlier by Singer (1936) and Allen (1954).

2 The successive curves of city sizes in the United States (fig. 2.3) demonstrate the point, see above, p. 30.

assume that there is even a very slight random fluctuation of individual city growth around this common rate, the effect is very different. Let us assume that this fluctuation conforms to Simon's (1955) growth model, so that the disturbance term is identical within any band of city sizes. This fluctuation can be of the following order: within each size band, half of the cities will stay the same size, one-quarter will increase by 10 per cent, and one-quarter will decline by 10 per cent.[1] The progression of sizes will then be as follows:

	Size of city							*Total number of cities*	*% of total population in top 10 cities*
Time	*7,110*	*7,900*	*9,000*	*10,000*	*11,000*	*12,100*	*13,310*		
t_1				128				128	7·81
t_2			32	64	32			128	8·59
t_3		8	32	48	32	8		128	9·28
t_4	2	12	30	40	30	12	2	128	11·53

The percentage of the total population which is found in the ten largest places increases progressively over time from 7·8 at time t_1, to 8·6 in t_2, 9·3 in t_3 and 11·5 in t_4. Despite the fact that there is a common overall rate of growth and that the fluctuation around that rate is the same for groups of larger and smaller places, the degree of concentration will increase. Only if there is *no* fluctuation whatsoever and a common rate of growth, will concentration not occur. The greater the degree of fluctuation the more rapidly will concentration take place. Interest in the regression formulae should therefore be focused upon the size of the disturbance term as well as upon the closeness of the b term to 1·0. Only some form of countervailing tendency in the growth of cities will prevent increasing urban concentration from occurring.

This phenomenon of concentration was discussed very early by Galton who recognized the need for some form of 'regression towards the mean'[2] if the growth of units was not to lead to growing concentration. In the study of business-firm growth, this requirement has been discussed by Kalecki (1945) and by Prais and Hart (1956) and a similar regression towards the mean has been recognized in the study of IQs of parents and children.[3] A somewhat parallel

[1] By using size bands, this growth process conforms to the weaker assumption discussed above, pp. 31–2. It means that it is perfectly possible for any given city to increase consistently over successive decades. The percentage growth rates used in this instance should be regarded as departures from the overall average growth rate which is taken as being zero.

[2] So as not to cause confusion between the statistical technique of linear regression and Galton's term 'regression', the latter will be referred to as 'regression to the mean'.

[3] On average, the children of parents with high IQ tend to have slightly lower IQs than their parents and children of parents will low IQ tend to have slightly higher IQs than their parents. The regression towards the mean level of IQ operates to prevent the distribution of IQs from becoming progressively distended.

phenomenon can be seen in the discussion of positive and negative feedback effects in regional economic development.[1]

The need for some retarding element in order to maintain concentration at a constant level has briefly been noted in the case of urban growth (E. N. Thomas 1967) and the suggestion of retardation in the growth rates of individual cities has been discussed most fully by Madden (1956b and 1958) and Williamson (1965b; Williamson and Swanson 1966). It is by no means clear, however, that one can as yet reconcile the apparent stability of size-distribution curves with the inevitable tendency towards increasing concentration which is a concomitant of the urban growth process of proportionate effect which has so often been assumed in studies of urbanization.

REGRESSION ANALYSIS OF NINETEENTH-CENTURY URBAN GROWTH

Bearing in mind the discussion about urban concentration, our regression formulation can now be re-expressed in the following terms:

$$X_t = a + (1 + c)X_{t-1} + \epsilon; \qquad [3.2]$$

where c is Galton's constant – the negative weighting which is necessary to retard the growth of large places relative to small places by an amount which will maintain the level of concentration in the distribution under consideration. Obviously, this is directly the equivalent of equation [3.1] since the c term equals the slope coefficient, b, minus 1·0. Equation [3.2] simply argues that, to produce a distribution whose component units grow while maintaining the overall stability of the distribution, the larger units have to grow more slowly than the smaller. The size of the negative weight, c, which is required for stability will depend on the size of the disturbance term, ϵ, since, as was shown above, concentration will tend to increase more quickly the larger the variance around the common growth.

Table 3.4 shows the regression coefficients of towns' opening and closing sizes for each of the eleven decadal periods between 1801-11 and 1901-11. Throughout, the natural logarithm of population sizes has been used in the calculations. There are two points of particular interest. First, it is apparent that the values of b, the slope coefficient, are in every case very close to 1·0 which suggests that the growth of these nineteenth-century cities is governed by an

[1] Hirschman (1958), Myrdal (1957) and others argue that economic inequality is the product of cumulative growth which is aided by 'backwash' effects within a national economy and that, only at a later stage, is the level of inequality dampened by 'spread' effects operating from growth centres as negative feedback replaces positive. In a study of a number of national economies over time, Williamson (1965a) supports this temporal progression. Some of the checks to cumulative growth which have been suggested for national growth centres could obviously be applied directly to the growth of large cities.

Table 3.4 *Regression of urban populations: closing size on opening size of towns at successive dates*

Decade	*No. of towns in calculation*	*Intercept (a)*	*Slope (b)*	*b − 1 (c)*	*Coeff. of determination* (R^2)	*Standard error of estimate* ($s_{X_t \cdot X_{t-1}}$)
1801–11	256	0·009	1·017	0·017	0·977	0·128
1811–21	308	−0·047	1·028	0·028	0·980	0·127
1821–31	361	−0·147	1·037	0·037	0·978	0·128
1831–41	412	−0·159	1·035	0·035	0·970	0·160
1841–51	465	−0·186	1·037	0·037	0·978	0·131
1851–61	521	−0·086	1·024	0·024	0·968	0·166
1861–71	580	0·112	1·006	0·006	0·959	0·200
1871–81	691	0·049	1·013	0·013	0·967	0·183
1881–91	750	−0·012	1·015	0·015	0·975	0·166
1891–1901	834	0·065	1·008	0·008	0·974	0·166
1901–1911	885	0·148	0·998	−0·002	0·978	0·159

Notes

1. All calculations are for $X_t = a + bX_{t-1} + \epsilon$, where X_t is population at date t.
2. Values are for natural logarithms of town populations.

approximately constant proportion irrespective of city size. Second, however, it is not without significance that, with only one exception, all of the values are slightly higher than 1·0. If we subtract 1·0 from each of the values of *b* (thus giving the value of *c*), it can be seen that it is only in the final decade that Galton's constant gives a negative weighting. Indeed, there would appear to be a pattern of some regularity in the changes of the value of *c*. In the early decades it tends to increase in size, but in 1861-71 it falls markedly and stays at a low level until it actually falls to a negative value in the very last decade. If anything, therefore, the regression model of closing size on opening size suggests a weak positive relationship with the size of city for the early decades of the century, but that in the final quarter of the century and the first decade of the twentieth

Table 3.5 *Concentration of the urban population: percentage in the largest towns*

Date	*A* Total no. of towns used in calculation	*B* % of total urban population in largest 10% of towns	*C* Rate of change
1801	250	61·63	1·006
1811	300	62·02	1·013
1821	360	62·81	1·024
1831	410	64·31	1·020
1841	460	65·57	1·020
1851	520	66·88	1·017
1861	580	68·02	0·994
1871	690	67·58	1·015
1881	740	68·37	1·012
1891	830	69·18	1·000
1901	880	69·18	0·987
1911	920	68·28	

Note: Column C gives the ratio of the percentages (column B) at time t to those at time $t + 1$.

century this positive relationship was weaker and turned to a negative relationship in the decade immediately before the First World War. In all cases however, the relationship between size and growth rate is never strong.

The results of the regression approach might therefore suggest two conclusions. First that, at this aggregate level, the assumption that urban growth is a random proportion of existing population size does not appear to be unreasonable. Second that, since there is a marked disturbance term throughout and the variance of individual cities' growth rates is not compensated for by a negative *c* term which would produce 'regression to the mean', one would expect that the level of population concentration would have increased for all except possibly the final decade. The extent to which such growing concentration occurred during the nineteenth century is suggested in tables 3.5 and 3.6 which show, for

Table 3.6 *Concentration of the urban population: dispersion of sizes*

		Starting year			*Closing year*		
Set of towns	*No.*	*Mean* ($\bar{X}$)	*Standard deviation(s)*	*Coefficient of variation* (V)	*Coefficient of variation* (V)	*Standard deviation(s)*	*Mean* ($\bar{X}$)
1801–11	256	8·608	0·801	9·308	9·408	0·824	8·763
1811–21	308	8·623	0·813	9·434	9·583	0·845	8·819
1821–31	361	8·694	0·840	9·665	9·935	0·881	8·872
1831–41	412	8·768	0·878	10·012	10·345	0·922	8·915
1841–51	465	8·816	0·915	10·373	10·708	0·959	8·957
1851–61	521	8·886	0·947	10·661	10·939	0·986	9·011
1861–71	580	8·923	0·976	10·939	11·030	1·002	9·086
1871–81	691	8·977	0·979	10·903	11·027	1·008	9·141
1881–91	750	9·061	1·014	11·187	11·342	1·042	9·188
1891–1901	834	9·096	1·037	11·398	11·468	1·059	9·236
1901–1911	885	9·170	1·065	11·610	11·553	1·074	9·299

Notes

1. Each set of towns is comprised of those which had populations $\geqslant$ 2,500 at both the start and close of the decade shown.
2. Values are for natural logarithms of town populations.

the whole period of 110 years, both the percentage of the total urban population which was contained in the top 10 per cent of the cities as well as the standard deviations of the whole array of cities. It is evident that both measures suggest similar conclusions; that the rate of concentration increased in the period up to 1861, but thereafter slowed down somewhat and that concentration decreased in the final part of the whole period.

We might try to draw together some of these apparently disparate forays on the body of data by developing the ideas inherent in the regression model. The results suggest that equation [3.2] provides a useful model by which to represent the growth of cities over successive dates. It argues that the population of a city at one date is some common proportion of its size at an earlier time irrespective of that size, but that there is a (log)normally-distributed fluctuation in the growth rates of individual cities which can be seen as a disturbance term common to the whole set of places. The empirical data which fit this model therefore provide, for nineteenth-century England and Wales, some support for the idea of a form of stochastic growth process which has been the underpinning of many of the theoretical models of city-size distributions. However, this is to ignore the generally positive value of the c term which suggests that there is a deviation away from constant growth to one with a slightly positive relationship with the previous size of a city. This, and the variation in individual growth rates which is given by the ϵ term, suggest that the particular growth process operating in the nineteenth century would not lead to a stationary size distribution of cities, but rather one characterized by increasing concentration of population. Ideally, it would be of some interest to calculate the value of the c term which, for a given value of ϵ, would produce a distribution which would become neither more nor less concentrated over time. We know that a positive c term will lead to greater concentration, but how large a negative term would be needed to halt the progress of concentration? The calculation of such a value would be very easy were one dealing with a constant set of cities, but the fact that the set of places is increasing in number complicates its calculation unduly. Here is one of the further difficulties encountered in dealing both with the 'birth' of new places and the 'death' of declining places. It will be recalled that the second of Simon's assumptions about growth was that new places (or, in his case, new words) were added at a constant rate and that it was this rate which determined the closeness of the convergence parameter of b to a value of 1·0 in the growth formula noted earlier.[1] The rate of addition of new places in nineteenth-century England and Wales, however, was far from constant. Table 3.7 shows the 'births' of towns as a percentage of the total number of cities at the start of each decade and shows that this percentage declined from a 'high' of over 20 down to almost 12 in 1851-61, then rose markedly almost to 20 once again in 1861-71, but thereafter fell drastically to figures of less than 10 in all but one of the following decades

[1] See above, p. 32.

and ended with a 'low' of under 5 in 1901-11. Such varying rates of addition to the set of places complicates our assessment of Galton's constant – the value of *c* which would be required to maintain a stable distribution. For decades when the rate of entry of new places is high (assuming, as is the case, that the majority of 'births' was in the small size groups), the effect will be to increase the rate of concentration of population, whether measured in terms of the percentage of population found in the largest 10 per cent of the cities or in terms of dispersion. The opposite will apply when 'births' are low so that the markedly lower ratio of 'births' in the last quarter of the period would have the effect of increasing the concentration to a lesser extent so that the declining rates of

Table 3.7 *The rate of addition of new towns*

Decade	*A* *Total no. of towns at start of decade*	*B* *No. of new towns at end of decade*	*C* *B as % of A*
1801–11	256	53	20·70
1811–21	308	53	17·21
1821–31	361	51	14·13
1831–41	412	54	13·11
1841–51	465	57	12·26
1851–61	521	63	12·09
1861–71	580	113	19·48
1871–81	691	59	8·54
1881–91	750	86	11·47
1891–1901	834	59	7·07
1901–1911	885	42	4·75

concentration in that period must be seen as the product both of the falling value of *c* and of the falling ratio of 'births'.[1]

There are therefore three types of function which influence the extent to which a size curve will be stable or will show increasing concentration over time. Where the variance of the growth rates, ϵ, is high, concentration will proceed more rapidly; this concentration will be countered or exacerbated depending on the value of *c* – only where *c* is of a 'sufficiently' large negative value will concentration tend to be reversed; but in addition the rate of entry of new small towns has to be considered since it will tend to distend the tail of the lower end of the size distribution and it will do this to a greater extent when the rate is high than when it is low, with the consequent effect of increasing the speed of concentration.

[1] Fuguitt (1965) draws a parallel conclusion in suggesting that much of the apparent decline of very small communities in the United States is more a product of the declining rate of incorporation of new small places than of the actual decline of existing small places.

A DISAGGREGATED APPROACH

So far the data on urban size and growth rates have been studied in aggregate terms. It would be profitable to look in more detail at the growth rates of towns by analyzing not the whole set of places, but sub-sets of the whole population. This might throw more detailed light on the nature of urban growth in nineteenth-century England and Wales.

So that the individual growth rates of towns can be more readily comprehended, the actual populations, rather than the natural logarithms of populations, will be used. Further, in estimating the actual growth rates of individual towns, a slightly more complex measure of growth has been used than is normally the case. In the overwhelming majority of the work which has studied growth rates, a simple rate of population increase has been calculated for each decade, given by the formula:

$$G_{t-1.t} = (P_t - P_{t-1}).P_{t-1}^{-1}; \qquad [3.3]$$

where $G_{t-1.t}$ is the growth rate between the times $t - 1$ and t; and P is population. A more realistic estimate of the rate of growth and one which, as Gibbs (1961, p. 108) suggests, is a close approximation to a compound growth rate, is given by expressing the denominator of the equation not as the initial population of a place at $t - 1$, but as the mid-point population between the opening and closing dates. The calculation of growth is then given by the following:

$$G_{t-1.t} = (P_t - P_{t-1}).\left(\frac{P_t + P_{t-1}}{2}\right)^{-1} \qquad [3.4]$$

The use of this estimate of growth rates would seem to have two advantages. First, it is a more accurate assessment of growth since population change is a continuous not a discrete phenomenon and the estimated mid-point population is a closer approximation to the changing base from which increments to or decrements from the population occur during the course of a decade. Second, its use has the effect of altering the distribution of growth rates in such a way as to help counter the tendency of simple growth rates to be positively skewed.[1] This tendency has been widely noted (Madden 1956a; Lampard 1968, p. 128), but the fact that studies have produced sets of growth rates which have a long tail of very high values is partly the result of using simple rather than compound measures of growth. By comparison with a simple rate, equation [3.4] tends to compress the high positive growth rates and expand the negative rates. For

[1] It was partly to correct the different and much more extreme skew of town sizes that natural logarithms of population size were used in the regression analysis above, pp. 65 *et seq.*

example, a town which increased in population from 10,000 to 15,000 over the space of a decade would be considered, by using the simple growth rate of equation [3.3], to have a growth rate of 50 per cent; if it were to decrease in population to a figure of 5,000, it would be given a growth rate of –50 per cent. However, by using equation [3.4] which would involve denominators of 12,500 and 7,500 for the case of growth and decline respectively, its growth rates would be considered as 40 per cent in the case of expansion and –111 per cent in the case of decline. The tendency for positive skew in the distribution of growth rates is therefore counteracted to some extent by the use of this second estimate of growth and the assumption of normality which underlies parametric statistical tests will be more likely to hold true. The growth rates used in the following analysis are therefore all based on the somewhat modified form of equation [3.4] which approximates the compound rather than the simple growth-rate formula.

Growth rates for each town were calculated during each pair of successive census dates in which the town had a known population of over 2,500. The numbers of observations at each intercensal period are therefore comparable with the numbers used in the regression study: for the first decade there are 256 observations; for the final decade there are 885. The scatters of observations for each of the eleven decades are shown in fig. 3.2 which plots the growth rate against the population size of places on semi-logarithmic scales.

On casual inspection of these diagrams, the first clear impression is of the increasing number of observations during the whole period; the galaxies of dots become much denser in the later decades as more and more towns are 'born' into the population. Second, if one squints sufficiently hard, discernible patterns can be seen. It is clearly true, as the earlier regression analysis suggested, that growth and size are virtually unrelated. If one visualizes lines representing the average growth rates across the whole span of city sizes during each decade, then the lines would be roughly horizontal, suggesting that the whole set of places can be characterized as having a common growth rate around which disturbance occurs. If these lines are horizontal, then their vertical shifts upwards and downwards from one decade to another represent differences in the overall growth rate during the different decades.[1] For example, the average growth-rate lines for the earlier and middle decades appear to be characterized by average growth lines somewhat higher than those of the later decades and thus the whole set of places must have grown on average by a greater amount during the earlier part of the period. A third comment on the diagrams, however, must be that it is quite apparent that the scatter of observations around the imaginary average growth

[1] The relationship between these scatters of growth rates on the one hand and the earlier regression analysis of equation [3.2] on the other is quite apparent. A horizontal line of average growth rates corresponds to a slope coefficient of $b = 1{\cdot}0$ in the regression of closing on opening population sizes. The scatter around the line of growth rates equals the disturbance term, ϵ.

lines is not constant with increasing size of town. The disturbance appears to grow distinctly less as one moves from smaller to larger towns.

These latter two points are obviously of some importance in relation to our ideas about the overall growth rates of sets of cities and deserve a more detailed examination than the simple inspection of scatter diagrams. They are of importance because of their bearing on the stochastic growth models which have been widely assumed to apply to urban growth. Gibrat's law – or the law of proportionate effect – would state, as has been seen, that growth would be a common proportion of existing size and, as a strict and necessary version of the law, that the error term around this proportionate growth would be normally distributed. We therefore need to test whether or not these two properties are met in the empirical data of nineteenth-century growth in England and Wales.

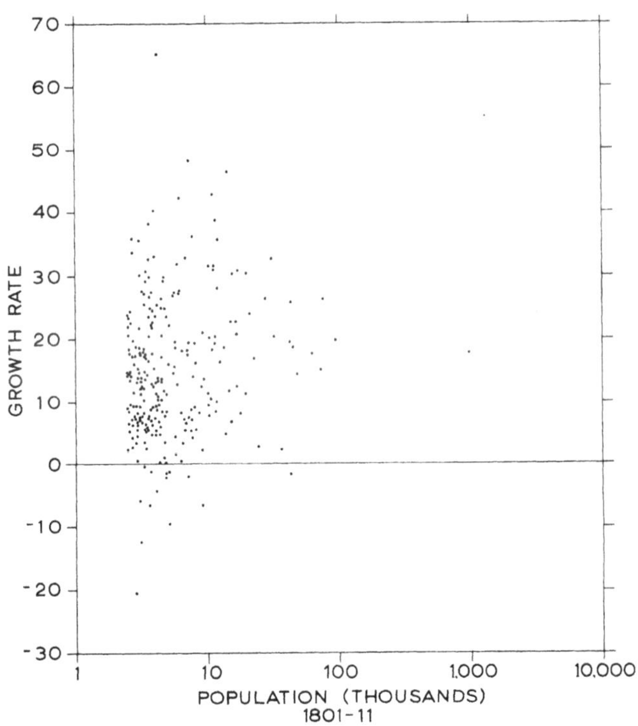

Figure 3.2 Growth rates and town size in England and Wales, 1801–1911. Note: A small number of places in the lower size ranges have growth rates greater than 70 per cent and are not shown in the scatter diagrams. The numbers involved, for each decade, are as follows: 1801–11, 1; 1811–21, 3; 1821–31, 3; 1831–41, 3; 1841–51, 3; 1851–61, 9; 1861–71, 13; 1871–81, 10; 1881–91, 6; 1891–1901, 11; 1901–11, 9.

continued

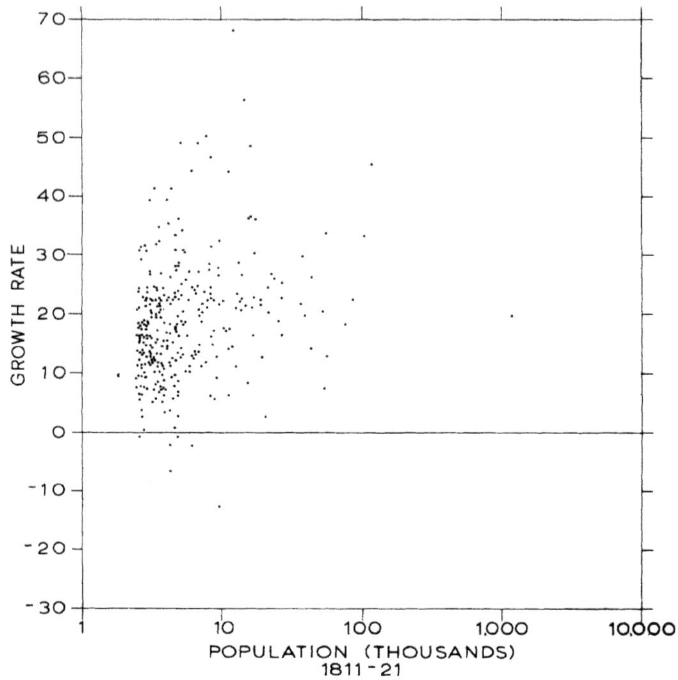

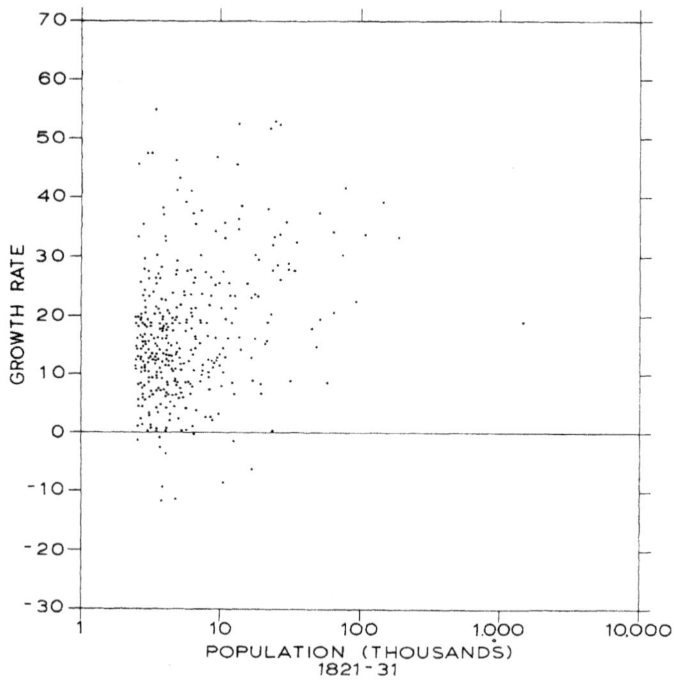

Figure 3.2 (*continued*)

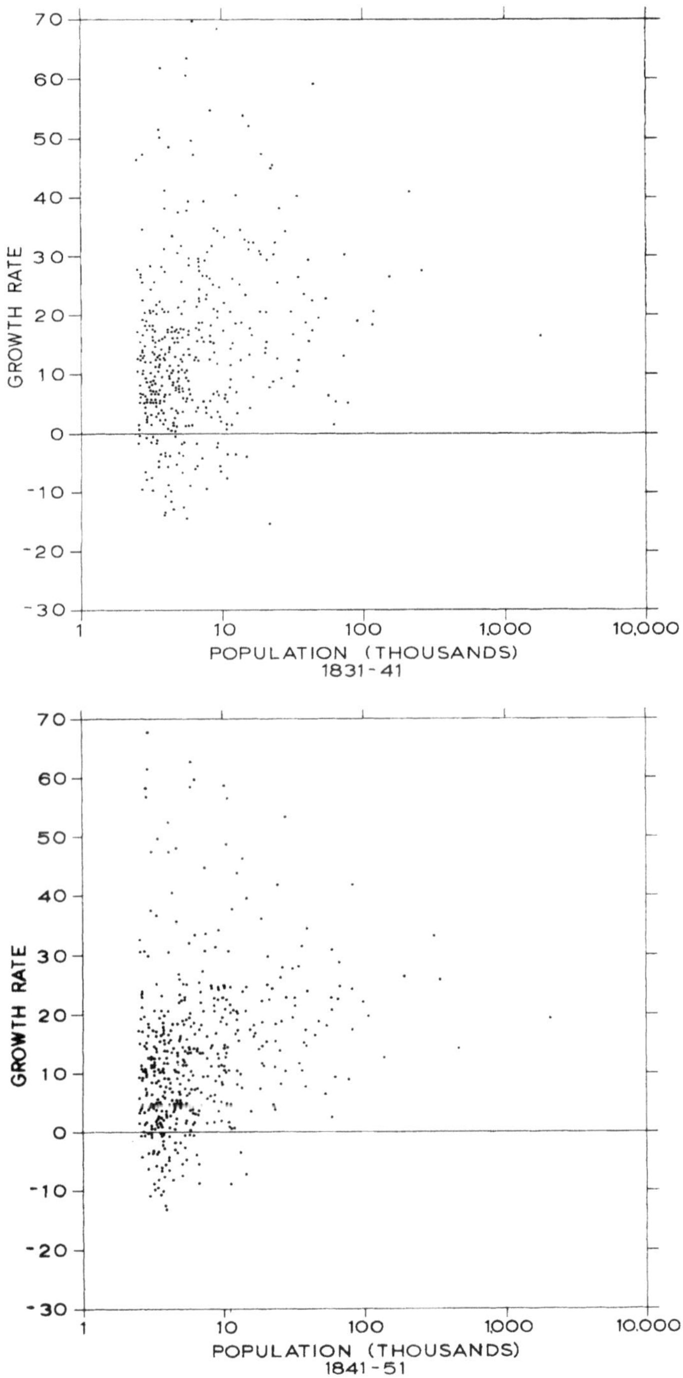

Figure 3.2 (*continued*)

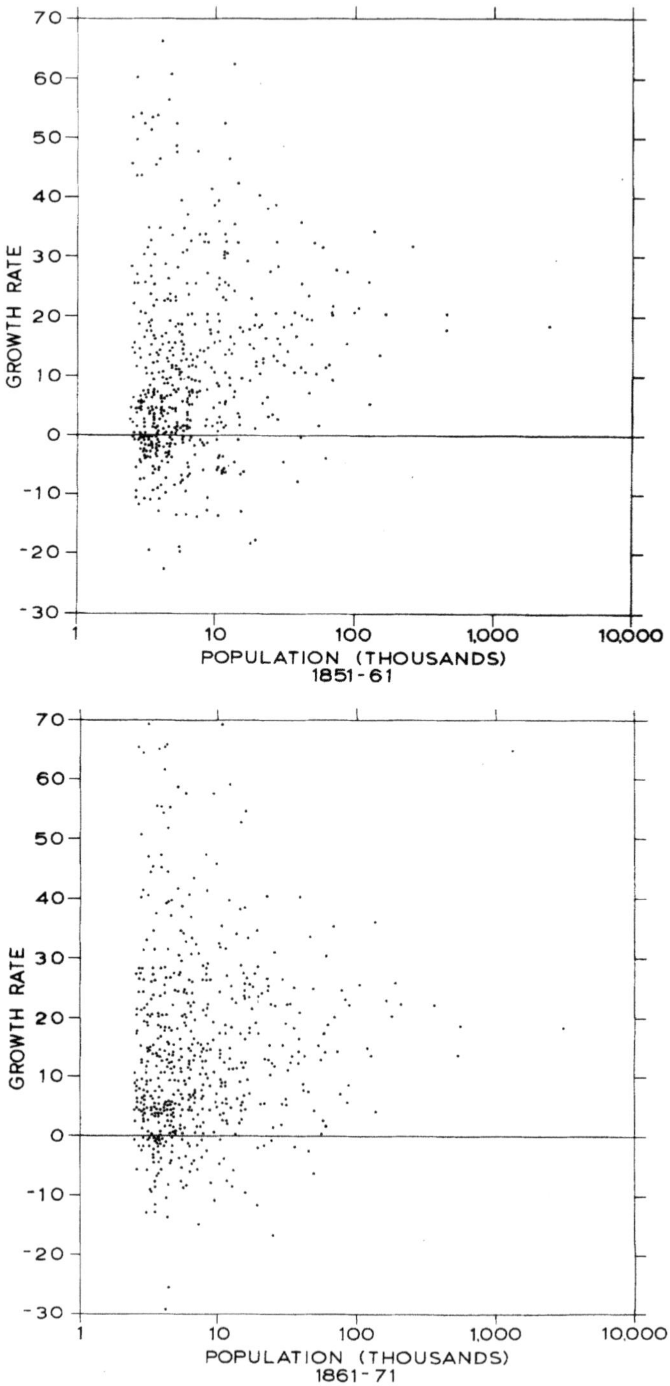

Figure 3.2 (*continued*)

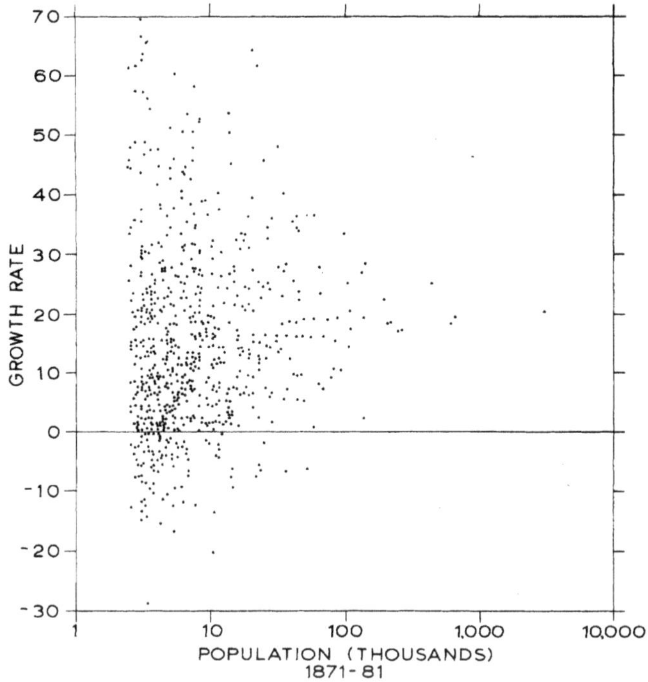

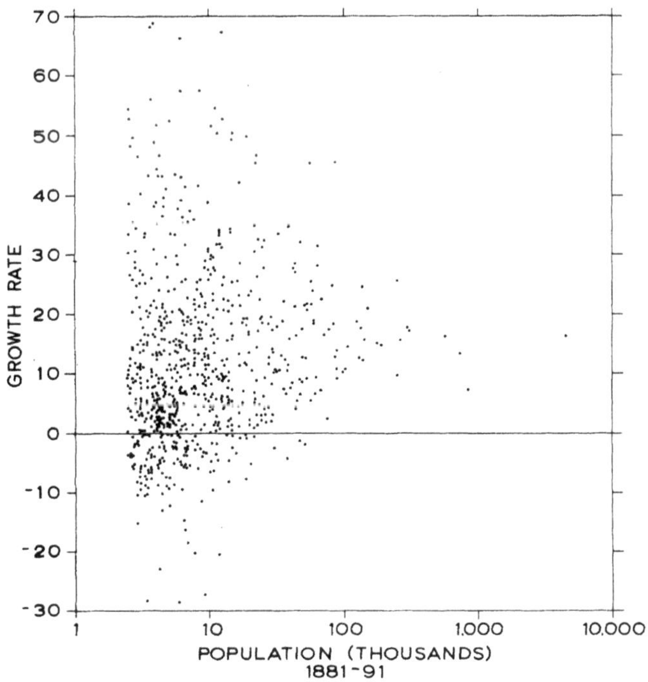

Figure 3.2 (*continued*)

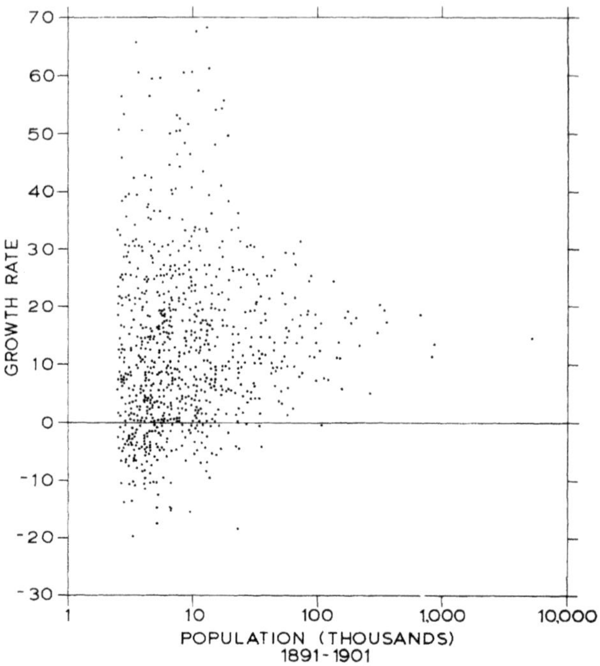

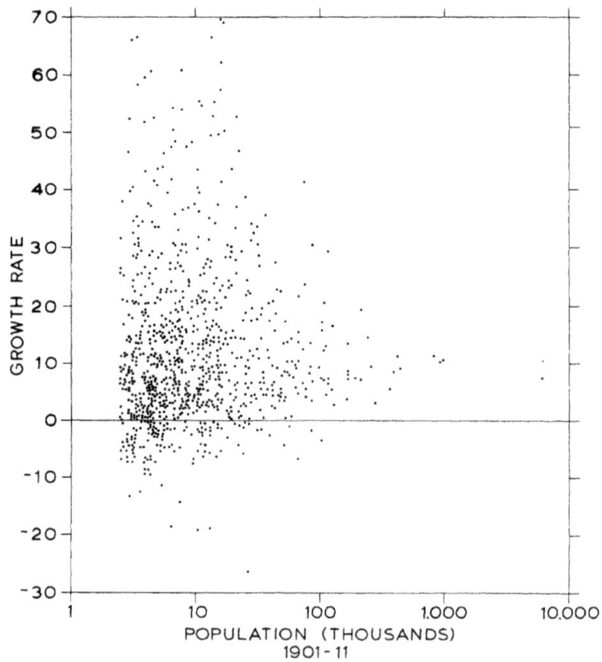

Figure 3.2 (*continued*)

One way in which this might be approached is, again, to use a regression model and table 3.8 gives the results of simple regressions of the following form:

$$G_{i.d} = a + b \log P_{i.t} + \epsilon; \qquad [3.5]$$

where G and P are respectively the growth rate and population of town i during the decade d which starts at time t.[1] The standard errors of these regressions formulations are, of course, very large, reflecting the expectedly low amount of explanation provided by the regression since the regression lines are virtually horizontal. Apart from the somewhat unreliable fact that the slopes of the relationship between size and growth rate are positive for all but the final decade, the regression approach is of little analytic value.

Table 3.8 *Regression of growth rates on town size*

Decade	*Intercept* (a)	*Slope* (b)	*Standard error of estimate* ($s_{G.P}$)
1801–11	0·71	3·94	12·11
1811–21	−4·63	6·43	11·67
1821–31	−14·63	8·53	12·59
1831–41	−15·80	7·97	15·49
1841–51	−18·59	8·49	13·81
1851–61	−8·31	5·35	17·09
1861–71	10·26	1·47	18·98
1871–81	4·28	3·06	17·58
1881–91	−1·83	3·67	15·91
1891–1901	7·25	1·72	16·45
1901–1911	14·95	−0·54	15·52

Notes: Values are for the equation $G_{i.d} = a + b \log P_{i.t} + \epsilon$, where $G_{i.d}$ is the growth rate of town i during decade d; and $P_{i.t}$ is the population of town i at the start of the decade.

A more profitable approach is to test the means and variances of growth rates for sub-groups of town sizes. Again, the size categories of 2,500-4,999, 5,000-9,999, and so forth, in which each group is twice that of the lower group, have been used. The values of the means and standard deviations for each of these size groups in each decade are given in table 3.9. The results are plotted in fig. 3.3 in which the means of each size group are joined by thick lines and thinner lines are drawn above and below them at one standard deviation from the means. These diagrams, and the data of table 3.9 on which they are based, are simplified descriptions of the galaxies of points showing the growth rates of each individual town.

[1] Equation [3.5] differs from the earlier regression equations of [3.1] and [3.2] in that it uses growth rate rather than closing population size as dependent variable. See above, pp. 62 and 65.

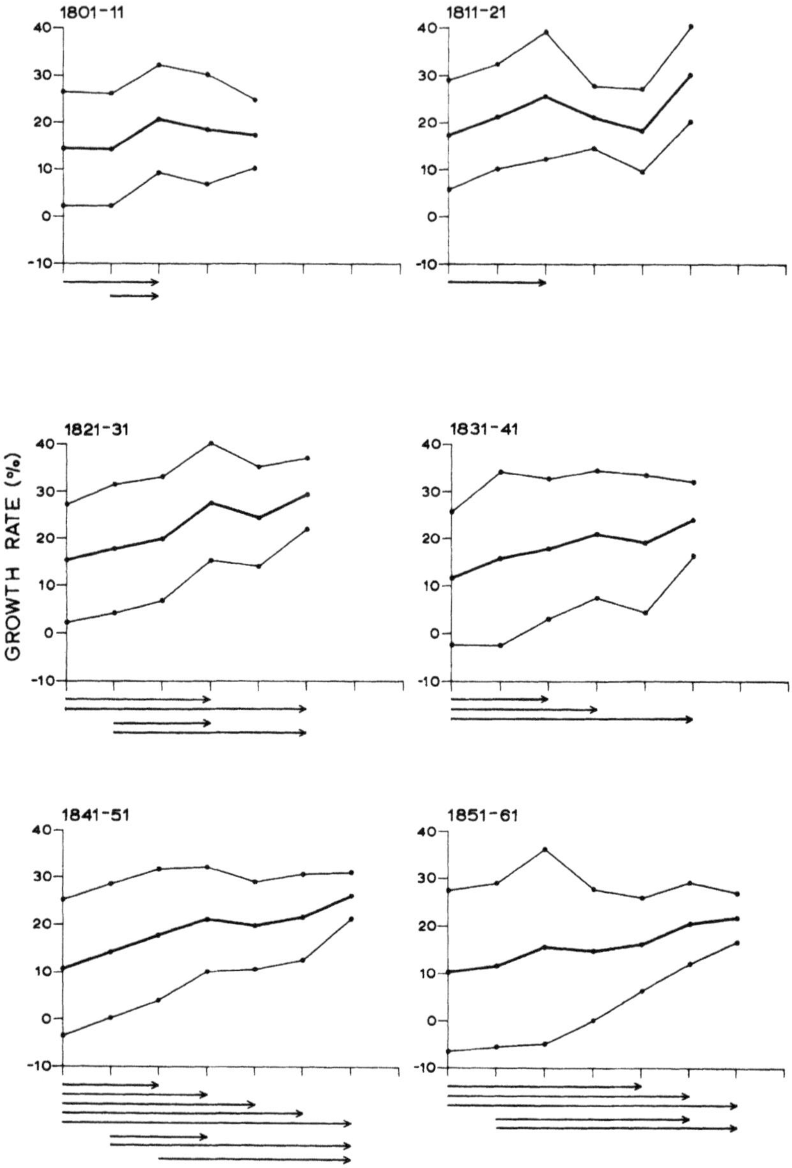

Figure 3.3 Growth rates and town size: means and dispersions. The diagrams summarize the data shown in fig. 3.2. Arrows below each plot show those size groups with mean growth rates significantly different at the 1·0 per cent confidence level, the arrow head points to that size group with the *higher* mean (see details given in table 3.10)

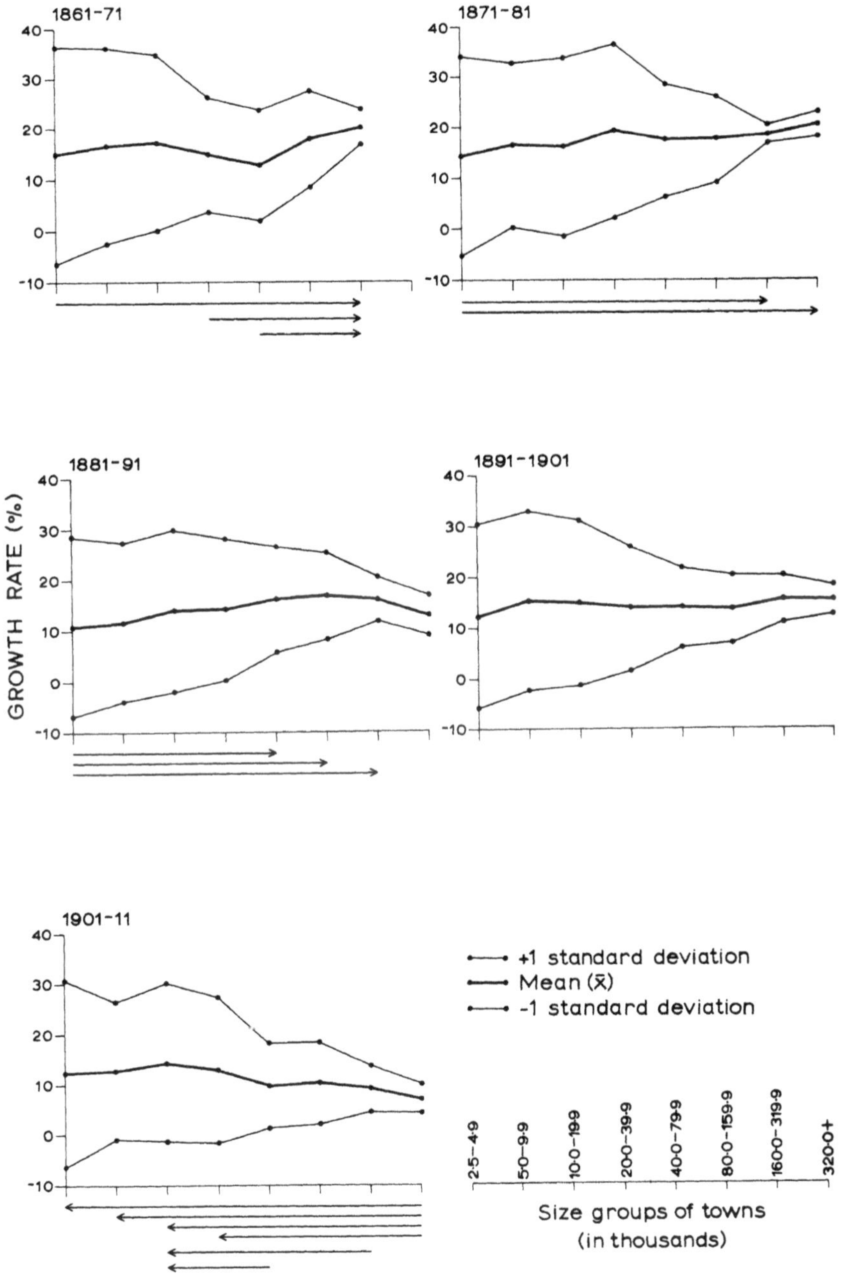

Figure 3.3 (*continued*)

Table 3.9 *Means and dispersions of growth rates of towns in different size groups*

Period		*Size group ('000s)* 2,500-	5,000-	10,000-	20,000-	40,000-	80,000-	160,000-	320,000-
1801–11	N	160	47	32	7	10			
	$\bar{X}$	14·5	14·3	20·7	18·6	17·5			
	s	12·3	12·1	11·6	11·6	7·4			
	$V\%$	85·0	84·9	55·8	62·5	42·2			
1811–21	N	191	64	31	12	7	4		
	$\bar{X}$	17·5	21·3	25·8	21·3	18·4	30·3		
	s	11·7	11·2	13·5	6·7	8·7	10·1		
	$V\%$	66·5	52·5	52·1	31·3	47·3	33·4		
1821–31	N	200	89	38	20	9	5		
	$\bar{X}$	15·3	17·9	20·0	27·8	24·6	29·6		
	s	12·0	13·6	13·1	12·4	10·5	7·5		
	$V\%$	78·6	75·9	65·5	44·5	42·7	25·5		
1831–41	N	207	115	44	26	13	7		
	$\bar{X}$	11·7	15·8	18·0	21·0	19·1	24·3		
	s	14·2	18·4	14·9	13·4	14·5	7·9		
	$V\%$	121·0	116·6	83·0	63·9	76·1	32·4		
1841–51	N	227	121	62	32	13	7	4	
	$\bar{X}$	10·8	14·3	17·9	21·1	19·9	21·8	26·2	
	s	14·3	14·2	13·8	10·9	9·1	9·0	4·9	
	$V\%$	132·0	99·1	77·3	51·7	45·6	41·4	18·7	
1851–61	N	231	135	85	31	23	8	5	
	$\bar{X}$	10·3	11·6	15·6	14·8	16·2	20·6	21·8	
	s	16·9	17·3	20·5	12·8	9·8	8·5	5·1	
	$V\%$	163·6	148·9	131·5	86·4	60·6	41·1	23·5	
1861–71	N	255	153	91	40	24	9	8	
	$\bar{X}$	15·0	16·9	17·5	15·1	13·0	18·4	20·6	
	s	21·3	19·2	17·4	11·3	11·0	9·5	3·5	
	$V\%$	141·7	113·7	99·3	74·4	84·1	51·5	17·0	
1871–81	N	274	201	111	55	28	13	5	4
	$\bar{X}$	14·6	16·8	16·5	19·8	17·8	18·1	18·9	21·0
	s	19·6	16·3	17·6	17·2	11·1	8·5	1·8	2·6
	$V\%$	133·8	97·0	106·8	86·8	62·2	47·0	9·6	12·2
1881–91	N	268	215	135	65	37	17	7	4
	$\bar{X}$	11·0	11·9	14·2	14·6	16·5	17·3	16·6	13·3
	s	17·8	15·6	15·9	13·8	10·4	8·5	4·4	3·7
	$V\%$	161·3	131·3	112·0	94·8	62·9	49·0	26·2	28·1
1891–1901	N	285	230	159	77	41	20	8	6
	$\bar{X}$	12·3	15·4	15·0	14·1	14·2	13·9	15·8	15·7
	s	18·2	17·7	16·2	12·2	7·7	6·6	4·5	2·9
	$V\%$	147·6	114·6	107·9	86·4	54·1	47·9	28·6	18·4
1901–1911	N	287	225	190	89	49	23	11	8
	$\bar{X}$	12·4	12·9	14·5	13·1	10·0	10·5	9·4	8·6
	s	18·5	13·6	15·8	14·5	8·5	8·2	4·6	2·6
	$V\%$	149·1	105·4	108·8	111·1	85·2	77·8	49·3	29·9

Notes: For each size group, observations are as follows: N = number of towns, $\bar{X}$ = mean growth rate, s = standard deviation of growth rates, $V\%$ = coefficient of variation (%) of growth rates.

In testing the two properties of Gibrat's law, the evidence on the relationship of variance and population size is unambiguous. In virtually every case, the standard deviations and the coefficients of variation grow progressively smaller for the larger size groups. While the progression is not necessarily smooth, there is a tendency throughout for there to be a gradual fall in the size of the variance with successively larger city-size groups. The assumption of uniformly-distributed error around the line of proportionate growth is clearly not met.

On the second Gibrat property – the invariance of the means of different size groups – the conclusion is less immediately apparent. The regression analyses have both suggested slight positive relationships between growth and size. Here, with more disaggregated material, we can use more discriminating tests of whether this is or is not significant. In making comparisons between the mean growth rates of the different size groups allowance has to be made for the fact that the size groups have different variances as well as different numbers of observations, and that the standard tests comparing means are consequently ruled out. However, an appropriate test which does take account of the different variances and numbers of the observations is that suggested by Aspin and Welch (1949; for its application to business firms, see Singh and Whittington 1968, pp. 75-80). The results of using the Aspin-Welch test in comparing the significance of the mean growth rates for each of the size categories are shown in table 3.10 which plots significant differences at the 5·0, 2·5, 1·0 and 0·5 per cent confidence levels. Taking the 2·5 per cent level, for example, it can be seen that, in the decade 1801-11, out of a possible total of ten comparisons between the means of size groups, only two are significantly different. In all of the eleven decades there is a total of 230 such comparisons in which no fewer than 58 show that one mean is significantly higher than another at the 2·5 per cent confidence level.[1] There is a marked pattern in the distribution of these significant differences from one decade to another. The earlier decades show a much larger number of significant differences than do the later decades and in all cases the pattern of differences is one of higher mean growth rates being found in the larger city sizes. In the later decades, there are relatively few such significant differences and in the very final decade a completely different pattern emerges with larger cities tending to have significantly *lower* average growth rates than smaller cities. This changing pattern is clearly seen in fig. 3.3 in which differences between means which are significant at the 1·0 per cent level are shown by sets of arrows for each of the decades.

This is a most interesting finding and one whose implications need to be spelled out both in terms of the theoretical notions of urban growth and in terms of the nature of the facts of the growth of towns in England and Wales in the nineteenth century.

[1] The exact pattern of significant differences would, of course, depend upon the particular size groupings whose means are chosen for comparison, but the overall tendency in these sets of sizes over the eleven decades is relatively clear.

Table 3.10 *Significant differences between mean growth rates in different size groups*

1801–11

	2	3	4	5
1	0	++++	0	0
2		+++	0	0
3			0	0
4				0

1811–21

	2	3	4	5	6
1	++	++++	+	++	++
2		0	0	0	0
3			0	–	0
4				0	0
5					+

1821–31

	2	3	4	5	6
1	0	++	++++	++	++++
2		0	++++	0	+++
3			++	0	++
4				0	0
5					0

1831–41

	2	3	4	5	6
1	++	+++	++++	+	++++
2		0	+	0	++
3			0	0	0
4				0	0
5					0

1841–51

	2	3	4	5	6	7
1	++	++++	++++	++++	+++	++++
2		+	++++	+	+	++++
3			0	0	0	+++
4				0	0	0
5					0	+
6						0

1851–61

	2	3	4	5	6	7
1	0	++	+	+++	++++	++++
2		0	0	+	+++	++++
3			0	0	0	+
4				0	0	++
5					0	+
6						0

1861–71

	2	3	4	5	6	7
1	0	0	0	0	0	++++
2		0	0	0	0	+
3			0	0	0	0
4				0	0	+++
5					0	++++
6						0

1871–81

	2	3	4	5	6	7	8
1	0	0	++	0	0	++++	++++
2		0	0	0	0	0	++
3			0	0	0	0	++
4				0	0	0	0
5					0	0	0
6						0	0
7							0

1881–91

	2	3	4	5	6	7	8
1	0	+	+	++++	+++	+++	0
2		0	0	++	++	++	0
3			0	0	0	0	0
4				0	0	0	0
5					0	0	0
6						0	0
7							0

1891–1901

	2	3	4	5	6	7	8
1	++	0	0	0	0	+	+
2		0	0	0	0	0	0
3			0	0	0	0	0
4				0	0	0	0
5					0	0	0
6						0	0
7							0

1901–1911

	2	3	4	5	6	7	8
1	0	0	0	0	0	−	−−−−
2		0	0	−−	0	−−	−−−−
3			0	−−−−	−	−−−−	−−−−
4				0	0	−	−−−−
5					0	0	0
6						0	0
7							0

Notes

1. Entries in matrices are as follows:
   - + } Mean of column size group is higher or lower than
   - − } mean of row group
   - ++++ significant at 0·5%
   - +++ significant at 1·0%
   - ++ significant at 2·5%
   - + significant at 5·0%
   - 0 not significant at 5·0%
2. Significance of differences is assessed by using values of

$$v = \frac{(y - \eta)}{\sqrt{(\lambda_1 s_1^2 + \lambda_2 s_2^2)}}$$

where $y = (\overline{X}_i - \overline{X}_{i+1})$ = difference between means of size groups: $\lambda_i = 1/N_i$, where N_i is number of towns in size group i; s_i^2 = variance of size group. See Aspin and Welch (1949) and supplementary tables in Pearson and Hartley (1966, table 11).

3. Size groups are as follows:

1	2,500– 4,999
2	5,000– 9,999
3	10,000– 19,999
4	20,000– 39,999
5	40,000– 79,999
6	80,000–159,999
7	160,000–319,999
8	320,000–

In the earlier decades, larger size groups have been merged to keep the number of towns in each cell ⩾ 4.

4. Details of N_i, $\overline{X}_i$ and s_i are given in table 3.9.

First, in terms of Gibrat's law of proportionate effect, both the decreasing variance of growth rates with size and this relationship between average growth rate and population size suggest that neither property of Gibrat's law are met.[1] During all the decades, variance decreases with city size and in only some of the later decades are the mean rates of growth of larger size groups not significantly higher than smaller sizes. This finding adds considerable weight to our earlier observation that the regression of growth on size tended to be positive. Now that the suggestion of the regression analysis is complemented by the more rigorous finding of significantly higher growth in larger city-size groups, it can be said that a uniform urban growth rate cannot be proven for most of the decades of nineteenth-century England and Wales. Were our tests of stability in the distribution of city sizes sufficiently powerful, the effects of this unequal growth could be demonstrated. The effect of higher average growth rates in larger towns would be to steepen the slope of plots of size against rank at successive decades as urban concentration progressed. If the average rate of growth grew progressively greater for larger towns, this would not necessarily have the effect of destroying an existing logarithmic-linear plot between rank and size, but it would certainly alter the overall shape of the distribution. In terms of an expectation that city sizes are distributed as the right-hand tail of a lognormal distribution, it would mean that the tail would become increasingly elongated as the degree of concentration of the population progressed. However, our second finding – that the variance of growth declines with city size – would tend to have a retarding effect upon this rate of concentration. To expand this point, one need only think back to the earlier discussion about the effect of the disturbance term in the regression formulae of equations [3.1] and [3.2]. The greater this term – the greater, that is, the variance in growth rates – the quicker

[1] No comparable formal tests of the applicability of both of the Gibrat properties to urban growth have been discovered. Thompson (1965b) has looked at the relationship of size and variance in his study of the growth rates of Standard Metropolitan Statistical Areas in the United States in the decade 1950-60 and finds the same pattern of declining variance amongst larger cities. His suggestion (Thompson 1965a, p. 22) on the relationship of growth and size is that, above a certain size of city, no cities decline in size and that there is a common growth rate. This he likens to an 'urban ratchett': 'at a certain range of urban scale, set by the degree of isolation of the urban place, the nature of its hinterland, the level of industrial development in the country, and various cultural factors, some growth mechanism, similar to a ratchett, comes into being, locking in past growth and preventing contraction'. The reasons adduced for this ratchett-effect are the diversification of industry, the existence of political power, the huge fixed investments, the rich local market and the steady supply of industrial leadership and innovativeness which characterize large urban economies. Thompson's later discussion of the relationship of cyclical instability and the size of a local economy (Thompson 1965a, p. 160) are also of some relevance here. Because of their industrial structure he hypothesizes that large places will approximate more closely to national cyclical patterns, but that smaller places will show greater variability since some have predominantly stable and others predominantly unstable industries. Studies of business-firm growth have tested Gibrat's assumptions much more extensively. Conclusions on firms' sizes, mean growth and variance of growth have differed somewhat both between and within studies in Britain and America (for a summary, see Singh and Whittington 1968, chapter 4).

will be the increase in concentration, assuming that the slope of the regression is held constant. In the regression model it was assumed that the disturbance term was both (log)normally distributed and was homoscedastic – in other words, that the variance was equal irrespective of size group. Since the disaggregated empirical data show that the variance is distinctly heteroscedastic and that it is the larger places which have smaller variance in their growth rates, the size of the disturbance term is lower for larger size groups and they will therefore tend to spread out in size less rapidly than will smaller places so that the overall rate of concentration of population will be less than would be suggested by the overall variance of the total set of places.[1] If this finding were to be fed back into the regression formulae to produce a model which might characterize the level of population concentration and the future sizes of sets of towns, the regression formulation would need to have added to it a vector of correcting factors given by the declining variances of larger groups of towns or it would need to be further partitioned into sets of equations of the following sort:

$$X_{t.s} = a + (1 + c) \,.\, X_{t-1.s} + \epsilon_s; \qquad [3.6]$$

where the subscript s is a size group of cities and the ϵ_s term would become smaller for increasing values of s. Obviously this is a far less attractive form of growth model since it is less general than equation [3.2] and because, without calibrating it for particular sets of data, it is impossible to use it to generate expected plots of future city-size distributions.

A more general suggestion from our findings, however, would be that there is great value in the more empirical historical study of sets of towns as against the more general theoretical growth models which draw on micro-economics and the behaviour of the firm. Lampard (1968, p. 95) suggests that there is a need 'for a distinctive *developmental* perspective on the evolution of city-regions, a perspective that allows for greater disaggregation and, at the same time, yields generalizations of a somewhat lower order than those of formal theory'; and even the economist Thompson (1965a, p. 60) says, 'It may be that a rigorous synthesis of many . . . case studies is needed to unlock the secrets of the urban growth process: the economic historian may have fully as much to contribute to urban growth theory as either the economic theorist or the econometrician'. It is this view which has buttressed the present concern with a more historical and a less formal systems' approach to the study of urban growth.

TWO QUESTIONS

Two interesting puzzles are posed by our findings on the empirical data of nineteenth-century urban growth. The pattern of urban growth which is outlined by the analysis of the scatter of means and variances calls first for the

[1] This would mean that any attempt at calculating the value of Galton's c at which concentration would not occur would be even more complex than suggested above, pp. 69–70.

development of ideas about how urban growth in a set of cities produces a pattern of declining variance with increasing city size and what factors might determine the varying slope of the relationship between growth rate and size. Second, since the relationship between growth rate and size appears to have changed during the course of the nineteenth century, how can this change be explained? Looking at the diagrams of average growth rates (fig. 3.3) and the results of the Aspin-Welch tests in table 3.10, the nature and timing of this change can be more precisely specified. There appear to be two types of pattern: on the one hand there is a scatter of urban growth in which average growth is positively related to city size; and on the other, a pattern in which growth and size are virtually unrelated so that the curve of average growth is not significantly different from horizontal.[1] In both cases the variance declines

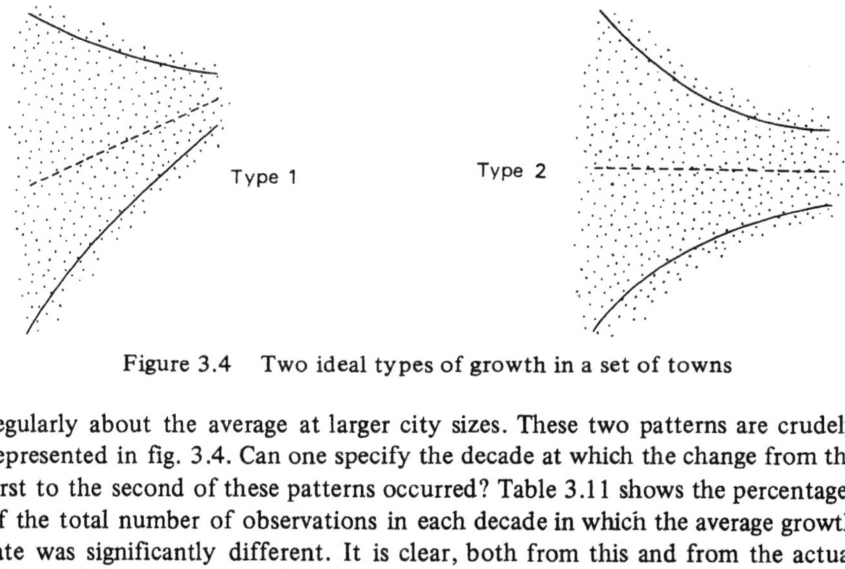

Figure 3.4 Two ideal types of growth in a set of towns

regularly about the average at larger city sizes. These two patterns are crudely represented in fig. 3.4. Can one specify the decade at which the change from the first to the second of these patterns occurred? Table 3.11 shows the percentages of the total number of observations in each decade in which the average growth rate was significantly different. It is clear, both from this and from the actual values of the means in table 3.9, that the proportion of significantly higher average growth rates is considerably lower in the decades after 1861 than in those before. The pattern of the first two decades is somewhat ambiguous, reflecting perhaps the greater imperfections in the census material, but they would both nevertheless appear to be closer to the first than the second of our ideal types. The third, fifth and sixth decades however conform very closely to our first ideal pattern in which growth is positively related to size; the fourth decade is less clear. The decades from the seventh onwards – that is, from 1861-71 – tend to grow increasingly close to the second pattern in which size

[1] These are, of course, idealized types. They ignore the pattern of the final decade in which the relationship is negative. Clearly, during the final quarter of the nineteenth century the once strongly positive relationship grew progressively less so until, by the final decade it had become negative.

and growth are virtually unrelated and this is especially clear in the decade 1891-1901. By 1901-11, the pattern has become the reverse of our first type with growth being lower for larger city sizes.

Table 3.11 *Percentage of comparisons between size groups in which larger towns have mean growth rates significantly higher than smaller towns*

	level of significance	
Decade	*1·0%*	*5·0%*
1801–11	20·0	20·0
1811–21	6·7	46·7
1821–31	26·7	53·3
1831–41	20·0	46·7
1841–51	38·1	61·9
1851–61	23·8	52·4
1861–71	14·3	19·0
1871–81	7·1	17·9
1881–91	10·7	28·6
1891–1901	0·0	10·7
1901–1911	21·4	39·3

Notes

1. Calculated from Aspin–Welch tests given in table 3.10.
2. For the final decade, 1901–11, the values show the percentage of comparisons in which large towns have significantly *lower* mean growth rates.

It is to discussion of these two questions that the bulk of the subsequent chapters is devoted. Chapters 5 and 6 develop a model of urban growth which attempts to simulate the declining scatter of variance around an average growth rate which can either be constant for the whole set of towns or can slope upwards to give a positive relationship between growth and size. Chapter 7 looks at certain aspects of late nineteenth-century Britain to throw light on the changing relationship of growth and size in England and Wales after 1861. But before turning to these, we must look at the spatial pattern of the growth rates which have been discussed in this chapter to see the extent to which general spatial patterns can be discerned which might throw light on the notion of sets of towns as formal systems with feedback loops and control mechanisms being reflected in the spatial pattern of their growth.

CHAPTER 4

Spatial patterns of urban growth

The eternal silence of these infinite spaces terrifies me. B. PASCAL

It is obvious from what has so far been said that the average rate of growth of a whole set of towns explains much of the expected growth of an individual town from within that set. Even when our model is disaggregated to produce average growth rates for differently sized sub-groups of that set, however, the model still suggests that there will be considerable deviation around the mean for individual cities. The growth model is a relatively weak one since, even though within a size group the fluctuation around the mean is random, for any individual town within that size group, growth from one decade to another can be consistently high or consistently low. The model, in other words, can accommodate the preferential growth of particular places during successive periods. The question then arises of how one explains the fluctuations of individual towns around the average growth rate of their size group.

In attempting to explain this scatter of individual growth, the most obvious approach for the geographer is to study its spatial pattern to see to what extent interpretable general features emerge which might throw light on the dispersion of growth rates. Does the existing literature on the spatial structure of economies suggest hypotheses which might be tested for the empirical data on nineteenth-century urban growth?

TWO SPATIAL HYPOTHESES

Two hypotheses may be proffered: first that, at a macroscopic scale, growth rates may fall into one of two patterns – either a single national pattern with one major core area of higher growth and a penumbral outer area with lower growth, or a regional pattern in which distinct regions show markedly different growth rates from each other and act as 'sub-systems' within the whole national area; second that, at a smaller scale, growth rates may decline with distance from regional metropolitan areas.

The literature from which such hypotheses are derived covers a wide span. Implicit in much of the geographical work on cities is the notion that, at successively higher levels of scale, functional urban regions 'nest' in to one another to form elements of an urban system at a higher scale. This of course is the skeletal argument of central-place theory which, while derived for areas much smaller than the whole national territory, has provided a major springboard for national studies. One of the earliest and most thorough of the national studies is Bogue's (1950) work on metropolitan dominance in the United States.[1] In it he demonstrated the way in which, at a macroscopic scale, a large variety of social and economic indices declined in value as one moved progressively away from the sixty-seven major metropolitan centres of the country. He argued that one could see the metropolitan centres as poles around which regional population distribution and types of economic activity were articulated and that, like the ecological community from which he drew his analogies, the constituent elements of the metropolitan community could be seen at a variety of territorial scales comprising 'dominant', 'subdominant', 'influent', and 'subinfluent' communities each of whose sphere of influence nested in to the next higher level. There have been many subsequent studies in the spirit of Bogue's, more especially within the United States, but, after the development of the concept of economic growth poles, such spatial analyses of national patterns of economic development have been extended to a wide variety of other countries both in the developed and developing worlds (for a geographical review see Keeble 1967).

Since the work on the United States has been the most fully developed, it is worth looking at it in some greater detail in order to exemplify the argument. The national economic structure which is suggested is one composed of an economic core region embracing the manufacturing belt which stretches from New York to Chicago and which 'performs' the role of national focus for the whole economy. Both within and beyond this core area, however, there is a large number of regional cores – or growth centres – which act as organizing poles, but at a regional rather than a national scale. Perloff's analysis is one of the most impressive (Perloff *et al.* 1960). It argues that the pattern of economic growth is more clearly seen if subdivided into two components which together make up the total shifts in regional employment: differential shift which is an intra-industry difference in the growth of regional employment by which, within any industry, certain regions grow more rapidly than others; and proportionality shift which is an inter-industry difference in which certain regions grow more rapidly because of their larger proportion of rapidly-growing industries. While the overall spatial pattern of total employment growth between 1939 and 1954 shows a wide scatter throughout the United States, when broken down into these two components the underlying spatial structure of growth becomes more

[1] The similarity between Bogue's ideas and central-place theory is striking. Bogue worked from the very different tradition of the human ecologists and, at the time, was unaware of Christaller's (1933) work which diffused to the English-speaking world very slowly, being first introduced to the United States in 1941 in a rather unexpected source, the *American Journal of Sociology* (Ullman 1941).

apparent. Differential shift is high within an essentially peripheral area with high growth being found in the fringe areas of the west, south-west and Florida, while within the manufacturing belt only the western edge shows high growth. Proportionality shift, however, shows a very different pattern with a distinct concentration of high growth within the manufacturing belt and only very small outliers in the fringe areas of Florida, southern California and Washington State. Perloff interprets these patterns in the following terms. The national core area of the long-established manufacturing belt is picked out by the measure of proportionality shift since its developed advantages of established infrastructure and access to the whole national economy foster within it the leading sectors of industry. The innovations associated with these new growth industries only filter out at a later date to the more peripheral parts of the country. The peripheral areas, on the other hand, grow predominantly on the basis of more localized access to raw material resources and also grow, again at a more local scale, on the basis of a process of infilling produced by the regional economy meeting successive demand thresholds which provide market opportunities for regional economic growth. The 'macro-geography' of the United States today suggests a much less clear emphasis on the national predominance of the manufacturing belt, but Perloff's analysis of the post-war period certainly accords with the idea of a single dominant national core area with regional sub-systems being subordinate to this core at a higher spatial scale. Exactly the same spatial pattern can be seen in the study *Metropolis and region* (Duncan *et al.* 1960) which looks at the structure of metropolitan centres of over 300,000 population in the United States. Here, it is the classification of metropolitan centres which reflects the same dichotomy that Perloff isolates. On the one hand there is a set of widely-spaced metropolitan centres which act, very much in the spirit of central-place notions, as centres which provide services for the localized regions which surround them. On the other hand, there is a set of much less regularly-spaced centres which are predominantly industrial and manufacturing complexes and which, largely concentrated within the manufacturing belt, rely on access to the market of the whole national territory.

For the United States one therefore has a body of well developed studies which points to the existence of an overall national economic system articulated around the twin principles of localized access to an immediate tributary area and generalized access to a national market. In very general terms, the economy can be conceived as a 'system' comprised at lower scales of 'sub-systems' which nest in to one another. The literature on the spatial patterning of urban growth has been able to call on certain of these ideas (Madden 1956b), but has also been able to isolate consistent spatial regularity in the growth of cities because of the particular sequence by which the country's urbanization proceeded. With the frontier of development being pushed progressively further west, the historical process of urban growth in the United States can be seen as a consistent sequence of western expansion followed by subsequent infilling and it has been this rather marked spatial pattern which has been most strongly emphasized by the geographical studies of urban growth (Luckermann 1966; Borchert 1967).

For Britain, there is not the same body of massive studies of the whole national urban system. However, the general outline of the spatial pattern of economic development is familiar. Looking at a longer span of time than just the nineteenth century, the changing importance of the national core area around London appears quite clearly. Before the industrial revolution, London's dominance of the national economy was unchallenged. Not only was population overwhelmingly concentrated in the South East – as Mantoux's (1928, pp. 350-3) rather charming maps reveal[1] – but the influence of the Metropolis extended over a very broad swathe of the country. Fisher (1935), for example, has shown the increasing extent of the area upon which London drew for food in the period from 1540-1640 and Wrigley (1967) has suggested the importance, in the following century, of the market represented by London on the production of foodstuffs and raw materials elsewhere in the economy and also, through the effects of migration, on changing the attitudes and horizons of a large proportion of the total population.[2] 'Town' to the man of the eighteenth century meant simply London – all the rest was 'Country' – and this contemporary evaluation was an accurate reflection of the marked difference not only between the population, but also the importance of the Metropolis by comparison with its nearest rivals. Even by 1801, there was an enormous differential between London – with its population of almost one million – and the largest provincial towns, none of which had reached 100,000. With the industrial developments of the nineteenth century, this spatial pattern of dominant Metropolis and subservient provinces was radically altered. While previously growth and activity had flourished in the South East – in what Sir Cyril Fox called 'Lowland Britain' which, certainly since Roman times, had enjoyed an almost uninterrupted monopoly of political power and economic wealth – the growth of industrial production was focused on the mineral-bearing and carboniferous areas of 'Highland Britain'. The pattern of demographic changes illustrates this radical alteration to good effect. The counties which showed a continuous rise in their proportion of the total population between

1 There are surprisingly few maps of population distribution in eighteenth- and nineteenth-century Britain. Petermann's splendidly detailed maps which accompanied the 1851 census are an outstanding exception. For periods before 1801, however, virtually no attempts to map regional distributions can be found. Those in Usher (1921, pp. 95-9) give a somewhat different impression of density than do Mantoux's.

2 Later developments in the sources of London's food supplies are suggested by Wetham (1964) who suggests that the radius from which milk was supplied widened rapidly at the end of the nineteenth century. Whereas in the 1860s milk 'warm from the cow' was supplied from stalls within London itself or from out-of-town farms at places such as Hendon, Dulwich and Highgate, by 1900 central London drew either directly on sources within thirty to fifty miles away or, through railway depots, on distant sources in Dorset, Wiltshire and Derbyshire.

1801, 1861 and 1921 were overwhelmingly those of 'Highland Britain': Northumberland and Durham, Yorkshire, Lancashire and Cheshire, Staffordshire, Worcestershire and Warwickshire and Monmouth and Glamorgan. In 'Lowland Britain', by contrast, only the metropolitan area itself and the area of the south-east coast and Hampshire basin continuously expanded their percentage of the total population throughout this period. The twentieth century, however, has seen a reversal back towards the older pattern of population concentration in the South East. Between 1921 and 1961, those areas which increased their proportional representation of population were all south of the Tees-Exe line; indeed, within that area, only Lincolnshire, East Anglia and the metropolis itself did not increase their percentage and in the case of the latter this was a reflection simply of the localized spill-over of population beyond the boundaries of London itself.[1]

Nineteenth-century developments can be seen therefore as something of an interlude in the longer-established pattern of metropolitan dominance. The expansion of the northern industrial areas, however, occurred in different regions of the country at somewhat different periods as the wave of growth industries progressed through the sequence of textiles, iron and the development of the railways, coal and, by the early years of the twentieth century, the motor car.[2] Largely on the basis of different factor endowments these various growth impulses had their greatest effect in rather specific areas of the country: cotton textiles in south-east Lancashire; woollen manufacture in West Yorkshire and Lancashire; iron in the West Midlands, Teesside and south Yorkshire; coal in South Wales and the North East. It is perhaps these industrial provincial nodes which might be seen as forming the cores of regional sub-systems in the development of the nineteenth-century economy. Briggs (1963), for example, suggests that one can characterize the developments in terms of the leading regional centres each of which developed sequentially as the 'leading cities' of their period: Manchester he calls the 'shock town' of the 1840s; its role was later assumed by the Yorkshire towns of Leeds and Bradford, each vying with the other in building bigger and more substantial municipal monuments to their provincial grandeur; subsequently Birmingham assumed the mantle of leading city, reaching its apogee under the civic gospel of Chamberlain in the 1870s; only by the 1880s or 1890s had London again begun to reassert its domination, a domination which differed from that of the provincial cities in that it was national in scope. Read (1964) makes a similar point, although suggesting different dating, in tracing the rising strength of provincial influence in the

[1] This description of regional percentages is based on Osborne (1964). Of the many studies of regional population distribution, an excellent cartographic portrait of the late nineteenth-century population changes is given by Lawton (1968b) who uses detail at the level of registration districts. For the early twentieth century, a useful study of the 1921-51 period is that by Willatts and Newson (1953).

[2] A simple cartographic sequence of the spread of urbanization, 1851-1951, is given by Friedlander (1970).

nineteenth century. He suggests that it was Manchester and Birmingham which were the two dominant cities, the lead having changed from the former to the latter in 1857. Briggs suggests that the successive rise of provincial centres might be traced out by the opening of local institutions such as Literary and Philosophical Societies[1] or by the development of provincial newspapers, whose subsequent decline in the twentieth century again reflected the re-establishment of London's national dominance. Kitson Clark (1962, p. 103) has suggested yet a further rough index of progress in suggesting that the medical enlightenment of various towns might be measured by the date of establishment of medical officers of health; Leeds having appointed one in 1866, Manchester in 1868, Birmingham in 1872 and Newcastle – 'normally in these matters a laggard' – in 1873. Indices such as this are, as Kitson Clark himself recognizes, only a crude and imperfect measure even of medical progress, let alone of the general importance of provincial centres. Nevertheless, it seems indisputable that a discernible sequence of regional development did occur during the nineteenth century with first one then another area feeling the full effects of the technical and industrial progress of the period.

POPULATION AND ECONOMIC GROWTH

It is in the light both of such empirical studies and of the more theoretical writings that one might hypothesize that the spatial pattern of urban growth rates might reflect the existence of sub-systems of regional cities by showing patterns of distinctly regional growth in the earlier part of the nineteenth century to be followed later by growth focused predominantly in the London area as inter-regional differences became subordinate to the dominance of the Metropolis. In the language of systems, one might say that the once distinctive regional sub-systems would eventually become less clearly evident as the whole national system developed around a single core area. In making such a suggestion one is of course using urban growth as a surrogate for the expansion of rates of general economic activity and one is assuming that higher rates of population growth are valid reflections of more general economic expansion. This may not be as unreasonable an assumption as may at first appear. Demographic growth is the product both of natural and migrational change and, even though it would appear that differences in fertility were more marked in nineteenth- than in twentieth-century Britain,[2] a number of grounds can be suggested as to why one

1 Most of which pre-date Victoria's accession. The sequence of foundation dates however is suggestive: Manchester 1781; Liverpool 1812; Leeds 1820; Sheffield 1822 (Briggs 1963, p. 44).

2 For the regional differences in fertility, valuable estimates of the disparities in the early part of the century are provided by Deane and Cole (1962, pp. 106-22). For the most recent period, the fall in regional differentials is well illustrated by a comparison of the patterns of fertility rates in 1911 and 1931, which shows not only the lower average, but more particularly the markedly smaller regional dispersion by the latter date.

can argue that, in the nineteenth century, migration acted to bring the distribution of population and economic growth into close harmony to a greater extent than it has in the twentieth century. First it can be argued that the characteristics of nineteenth-century industry made it less likely that population by itself would attract industry to it because of the greater proportion of heavy and capital-goods industries which were more tied to raw materials than to access to markets.[1] Second, it might be argued that at least one of the possible reasons for higher fertility in certain areas was the very proximity of employment opportunities represented by local industries.[2] Third, and most significant, one can argue that the impediments to migration were significantly fewer in the nineteenth than in the twentieth century. It is impossible to make accurate comparisons between the amounts of mobility in the two centuries since even the best estimate of nineteenth-century migration are far from perfect, applying as they do to such large areal units as counties and being measures of net rather than gross movements.[3] The imperfect evidence does suggest, however, that the rate of movement was significantly higher for the nineteenth century. Friedlander and Roshier's (1966) figures for movement between counties suggest an overall fall in the longer-distance net movements within the country from the last quarter of the century and, while much of this represents the substitution of emigration for internal migration, the differences between the nineteenth and twentieth centuries within Britain is marked. At a local level, Pritchard's (1972) study of movement within Leicester notes that the most striking difference between the pre- and post-First World War periods was the sheer fall in the rate of gross residential turnover produced both by movement within the town as well as movement to and from it.

Certainly a number of convincing explanations can be proffered in support of higher mobility in the nineteenth century. First the prevailing economic and demographic conditions of the period would favour high rates of movement.

1 Although it would be a gross exaggeration to say that industry was not attracted by the existence of a large potential labour force in an area. Obviously the relationship of industry and population is two-way. London itself – in absolute terms the largest industrial complex in nineteenth-century Britain – had few raw materials other than those imported from outside, although it could be argued that linkages with existing industries acted as 'raw materials' which attracted new industrial activity to the Metropolis. It would be true to say, however, that population by itself would attract economic growth *relatively* less in the nineteenth than the twentieth century (Wrigley 1965).

2 Deane and Cole (1962, pp. 114-16) suggest that seven of the ten counties with the highest rates of natural increase in the period 1701-1830 were Cumberland, Westmorland, the North and West Ridings of Yorkshire, Derbyshire, Staffordshire and Shropshire – a belt of counties all but enclosing the textile area of Lancashire and Cheshire. They comment, 'An obvious inference is that the rise of Lancashire stimulated the growth of population in the surrounding areas, and that part of the population increase was then syphoned off into Lancashire itself'.

3 Newton and Jeffery (1951) suggest that, for the period 1948-9, of the population changes in most of the areas studied, net migration was less than 5 per cent of the total gross migration of population.

Higher fertility and larger family size[1] increased the potential pressure for movement and the greater regional differentials in wage rates in a population much of which lived close to or below subsistence level meant that the economic incentives to mobility were high. But second, and of especial importance, are the differences in the housing structure of nineteenth- and twentieth-century England and Wales. The greater ease of moving house in the nineteenth century acted as enabling agent in translating the desire to move into high rates of mobility. The important difference lay in the structure of tenures. The nineteenth-century housing market was predominantly one of privately-rented accommodation and the proportion of owner-occupied housing was very small; thus reducing the legal and financial restraints upon moving. By contrast, the expansion of local authority housing after 1919 with the residence requirements which are needed to qualify families for tenure,[2] the relatively rapid expansion of the owner-occupied category, and the various decontrol-on-vacancy acts passed in relation to privately-rented furnished accommodation have all introduced dampers on twentieth-century mobility.

If this difference in the gross rates of mobility in the nineteenth and twentieth century existed, it lends considerable additional weight to the argument that migration effected a much more rapid and sensitive adjustment between population and employment opportunities in the nineteenth century[3] and so bolsters the assumption that population growth reflects economic growth and can be considered to be 'caused' by it. Even though most of the growth of towns was the direct product of natural increase rather than directly the consequence of migration,[4] it can be argued that people who are born in an area and opt to stay there are making a zero-distance migration decision which will not be entirely independent of the employment and economic opportunities of that area.[5] We can therefore draw on the wider body of theory about the spatial

1 Statistics of these differences are provided by Mitchell (1962) and are discussed in, for example, Carr-Saunders and Jones (1927) and Marsh (1958).

2 The lower rate of mobility amongst council tenants is discussed, for example, by Brooks (1967). Donnison (1967, p. 210) suggests, however, that they move as often as other tenure groups.

3 Saville (1963, p. 2) emphasizes the importance of the New Poor Law of 1834 as a major element in first freeing labour from the restrictions of the seventeenth- and eighteenth-century settlement laws. With the introduction of the poor law, he says, 'the labour market was now as efficiently adapted to the requirements of expanding industrialism as political legislation could achieve'.

4 Even in London – the recipient of so large a proportion of the total migrants within the country (C. T. Smith 1951) – total population grew by approximately equal contributions of natural increase and migration in the period 1851-91 (Shannon 1935). The importance of natural increase in the growth of provincial towns is shown by A. F. Weber (1899, p. 244). In the final part of the century, the contribution of migration was even lower. The increase in population in the Bristol area after 1871, for example, was almost wholly the consequence of natural change (Shannon and Grebenik 1943, p. 12). And Welton (1911, pp. 15-16) shows that for the decades 1881-91 and 1891-1901, even of the towns with the largest gains by migration, the proportions were above 20 per cent only in very few cases.

5 An argument of Ward's (1963) which has already been noted above, p. 33.

pattern of economic growth to derive spatial hypotheses to test against the patterns of population growth in Victorian towns.

THE MEASURE OF URBAN GROWTH

First, however, we need to derive an appropriate measure by which we can compare the growth rates of towns. From what has been seen already of the scatter of urban growth rates, it is apparent that direct estimates of growth are inappropriate because of the relationship between the size of a town and the expected variance of its growth rate. The scatter of growth rates is very wide for small places and relatively narrow for large places. Plotting the spatial incidence of growth rates would thus tend to emphasize small places at the expense of large; the probability of London having an abnormally high or low growth rate in any decade would be rather low, whereas the probability for, say, Penzance, would be very much greater. Some form of standardization is therefore needed which will take account of these differences. The most suitable is the use of z, or standard, scores. These express each town's growth rate as a deviation from the average growth rate of the size group to which it belongs in any decade. For example, table 3.9[1] shows that for the decade 1901-11 the means and standard deviations of growth rates in the largest size group of towns were 8·57 and 2·56 respectively and in the smallest size group they were 12·40 and 18·48. Thus, for the largest towns, places whose growth rates lay between the two limits of 11·13 and 6·01 would be within plus or minus one standard deviation of their size group average; whereas, for the smallest towns these upper and lower limits would be 30·89 and −6·09 respectively. To qualify as being 'unusually' rapidly or slowly growing, a town would therefore require different growth rates depending on the size of the town being considered.

The translation of individual town growth rates into standard scores accords with the urban growth model which has been discussed earlier since it takes account both of the varying means and varying disturbance terms which apply to groups of town sizes. The calculation follows the standard formula:

$$z_i = \frac{G_{i.n} - \bar{G}_n}{\sigma_n};$$

where $G_{i.n}$ is the growth rate of the ith town whose population places it within the nth size group of towns: G_n and σ_n are the mean and standard deviation of the nth size group. To take an example: Birmingham had a population which put it in the largest size group in 1901 and had a growth rate of 11·28 per cent in the decade 1901-11; Newquay, in Cornwall, lay in the lowest size group of towns and had a very much higher growth rate of 34·53 per cent. Despite these different absolute growth rates, however, it can be seen that, in relation to other

[1] See above, p. 82.

towns of similar size to themselves, both places stood in approximately the same relationship to each other. The standard scores of the two places are as follows:

$$z_{\text{Birmingham, 1901–11}} = \frac{11{\cdot}28 - 8{\cdot}57}{2{\cdot}56} = 1{\cdot}06$$

$$z_{\text{Newquay, 1901–11}} = \frac{34{\cdot}53 - 12{\cdot}40}{18{\cdot}48} = 1{\cdot}20$$

The growth rates of both places thus lie just over one standard deviation above the mean of their appropriate size group.

The derivation of these scores has been gone into in some little detail because it is important that, in studying the patterns of standardized growth rates, the nature of the statistics is clearly understood. Measured in this way, the fact that a town has a 'high' growth rate means, not that its absolute percentage rate of growth was high by comparison with the national average in the particular decade, but that it was high in relation to the scatter of growth rates of towns of comparable size. Thus each decade will have approximately the same proportion of places with 'high' and 'low' standard growth rates irrespective of the national average; likewise, even in decades in which the average growth of large places is higher than that of smaller places, not all of the larger places could appear as having 'high' standard growth rates since it is the deviations from the average of the size group rather than the whole set of places which are measured. There is a variety of base lines against which to measure 'high' growth rates. The one selected here has the great virtue of taking into account the varying scatter of growth within different size groups of cities and of conforming to our model of urban growth.

THE SPATIAL PATTERN OF STANDARD GROWTH RATES

The pattern of growth during the eleven decades from 1801 to 1911 is shown in figs. 4.1-4.11. For each decade, places with growth rates of above or below one standard deviation from the mean of their appropriate size group are distinguished from towns which grew or declined by less than this.[1] The maps produce a striking sequence which provides a fascinating commentary on the progress of urbanization during the nineteenth century. The patterns of relative growth rates, both for individual towns and for regional clusters of towns, tend strongly to confirm the well-known pattern of nineteenth-century economic

[1] In discussing the patterns of above- and below-average growth rates, places whose growth was over one standard deviation below the mean will be referred to as 'declining' places. The term is used in a *relative* sense. In certain cases – more especially with small places – the populations of such towns did show an absolute decline in numbers, but more usually they expanded absolutely even though at an 'abnormally' low rate. The numbers of places which showed absolute decline in any decade can be seen in the sequence of diagrams in fig. 3.2 (pp. 73–8).

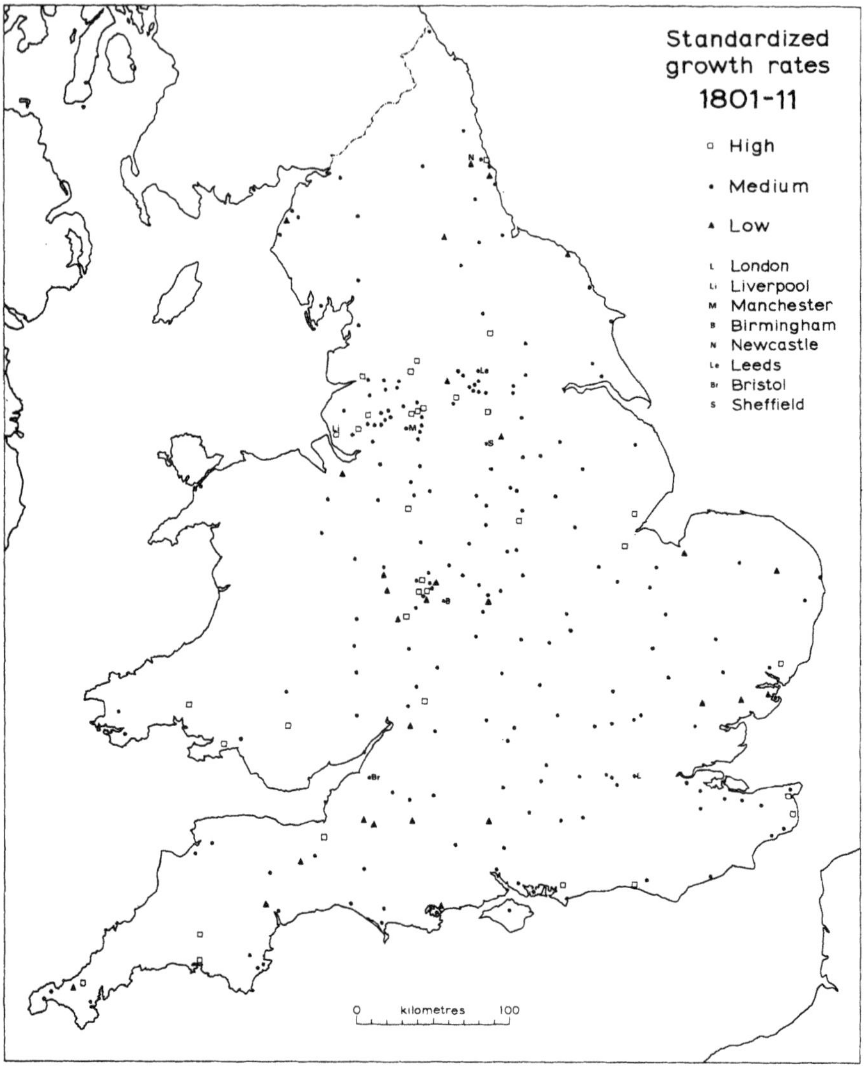

Figure 4.1 The spatial pattern of urban growth in England and Wales, 1801–11. Standardized growth rates are as follows: high = ⩾ 1 standard deviation above the mean for the appropriate size group; low = ⩽ 1 standard deviation below the mean for the appropriate size group; medium = between ±1 standard deviation from the mean of the appropriate size group. Details of means and standard deviations for each size group are given in table 3.9

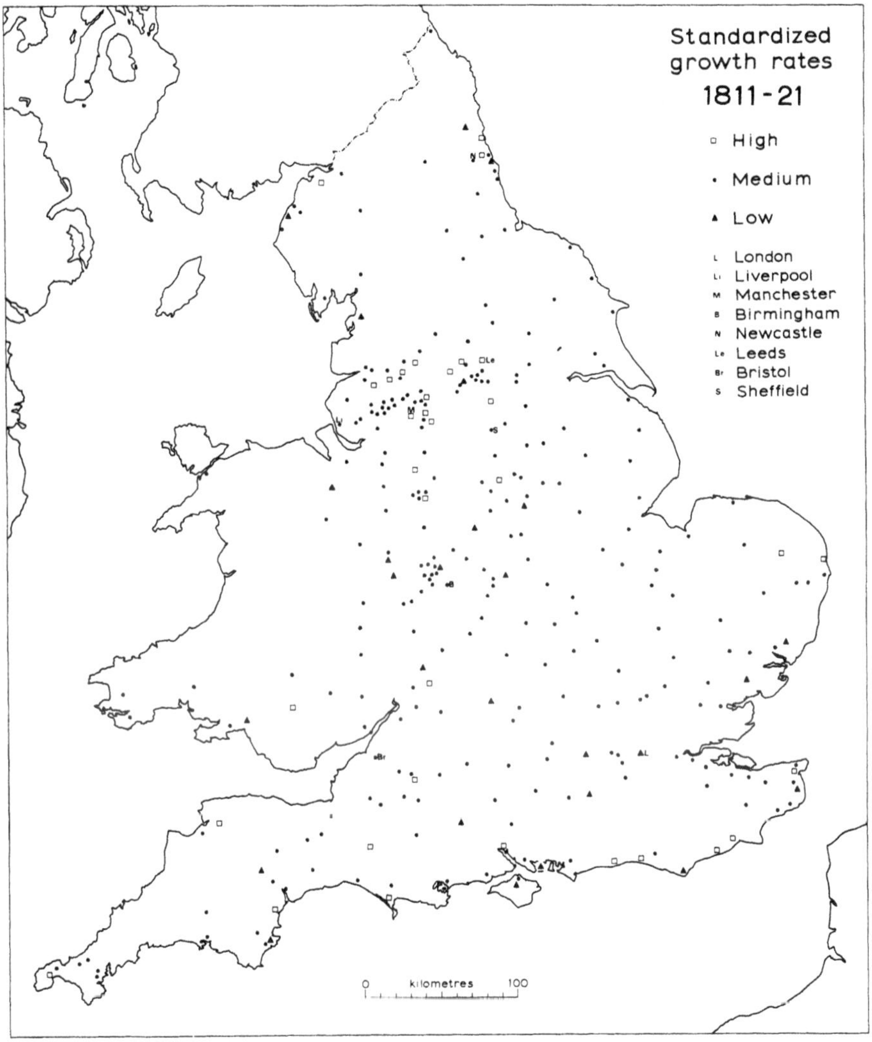

Figure 4.2 The spatial pattern of urban growth in England and Wales, 1811-21. For explanation, see fig. 4.1

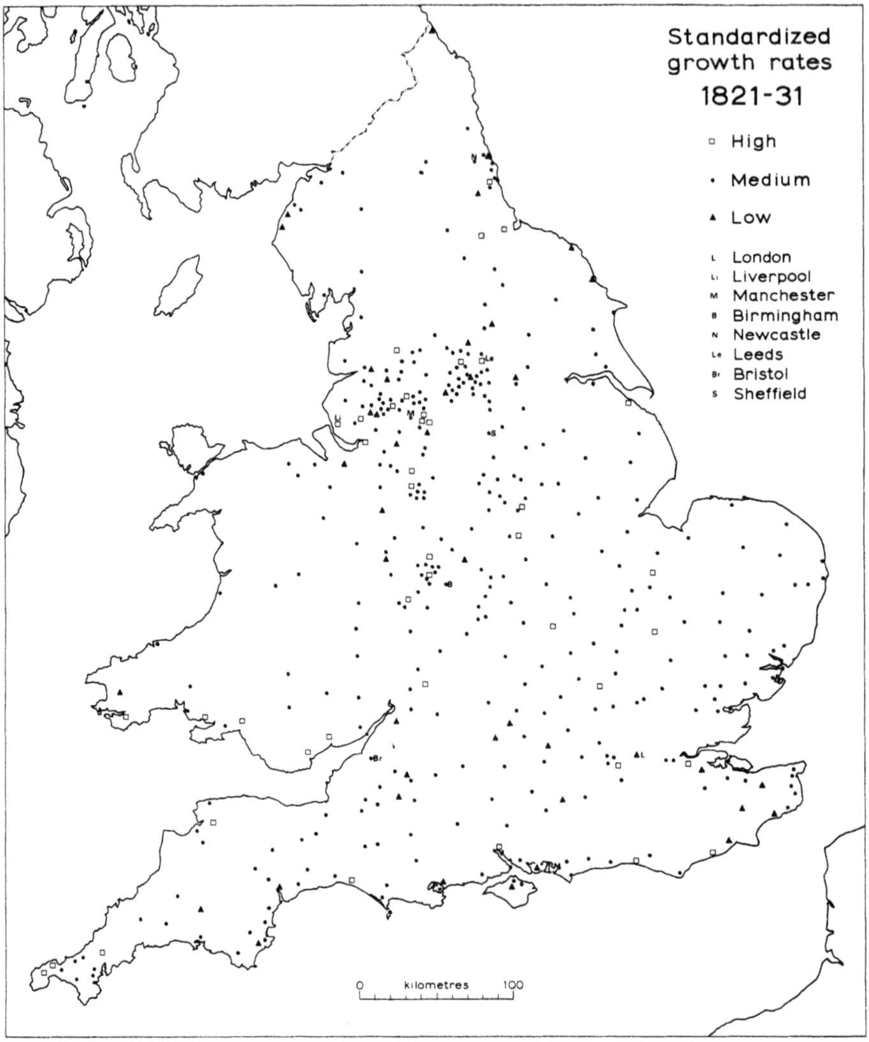

Figure 4.3 The spatial pattern of urban growth in England and Wales, 1821–31. For explanation, see fig. 4.1

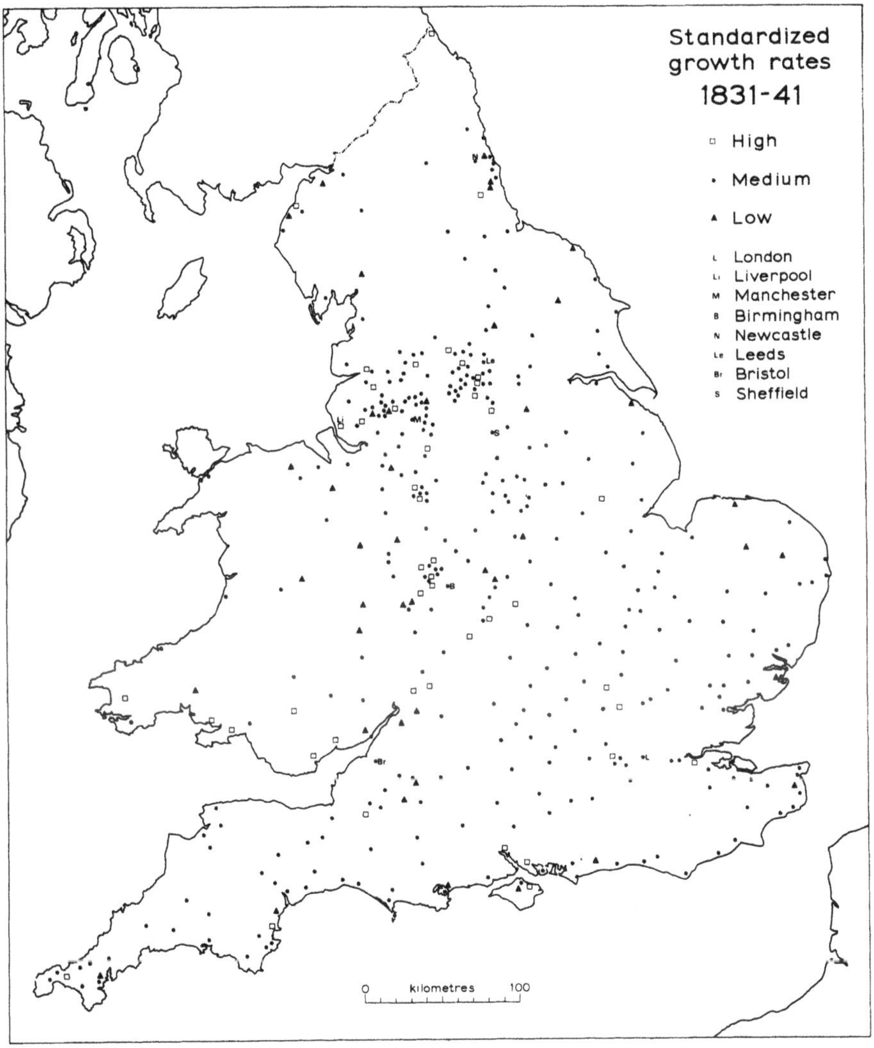

Figure 4.4 The spatial pattern of urban growth in England and Wales, 1831-41. For explanation, see fig. 4.1

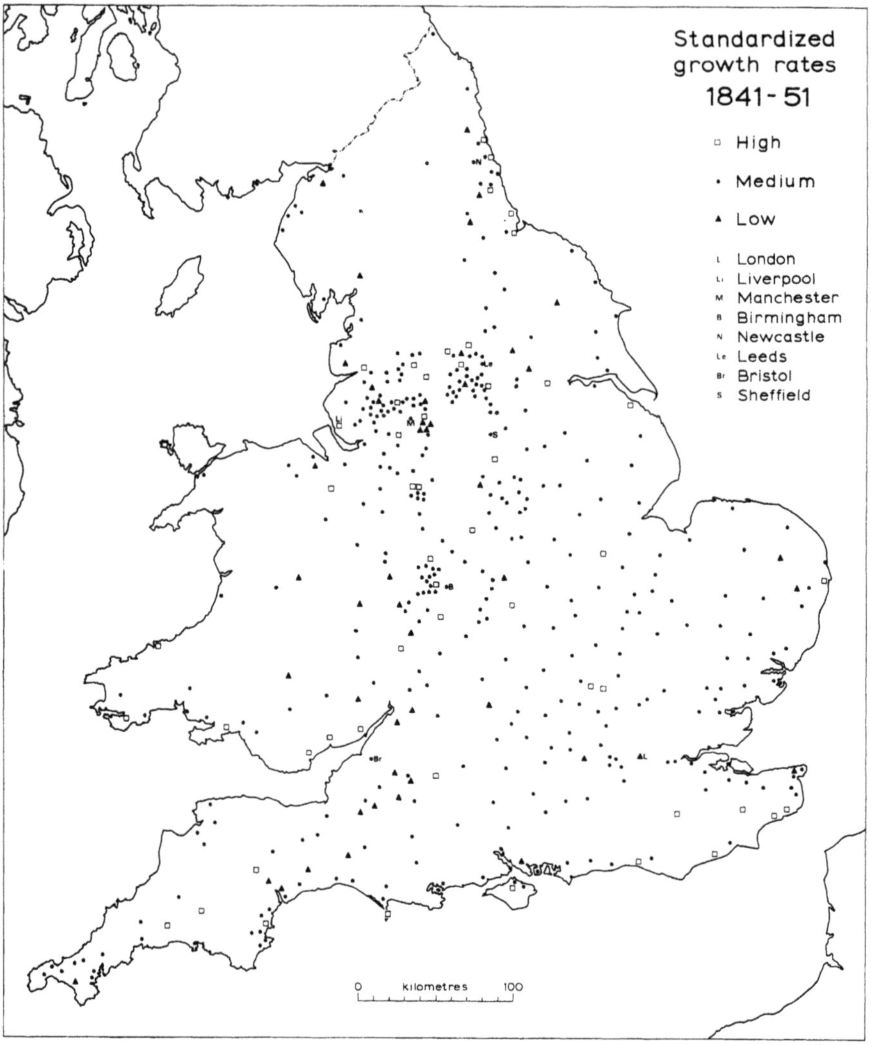

Figure 4.5 The spatial pattern of urban growth in England and Wales, 1841–51. For explanation, see fig. 4.1

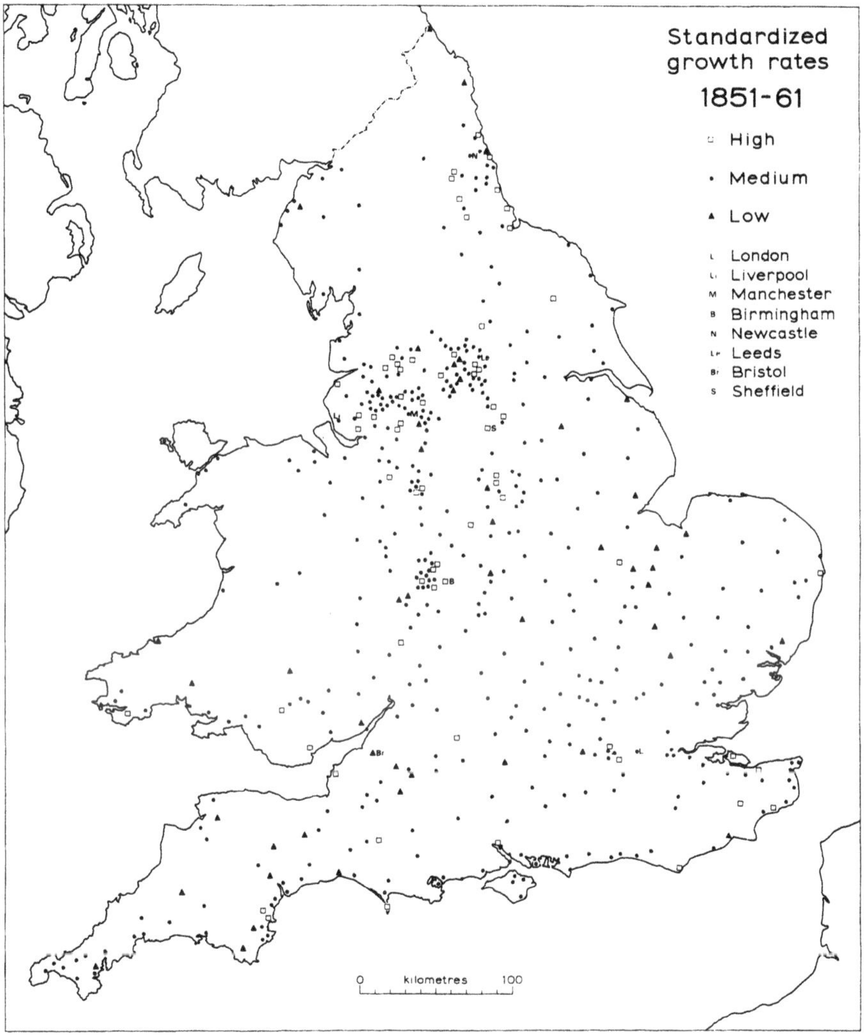

Figure 4.6 The spatial pattern of urban growth in England and Wales, 1851-61. For explanation, see fig. 4.1

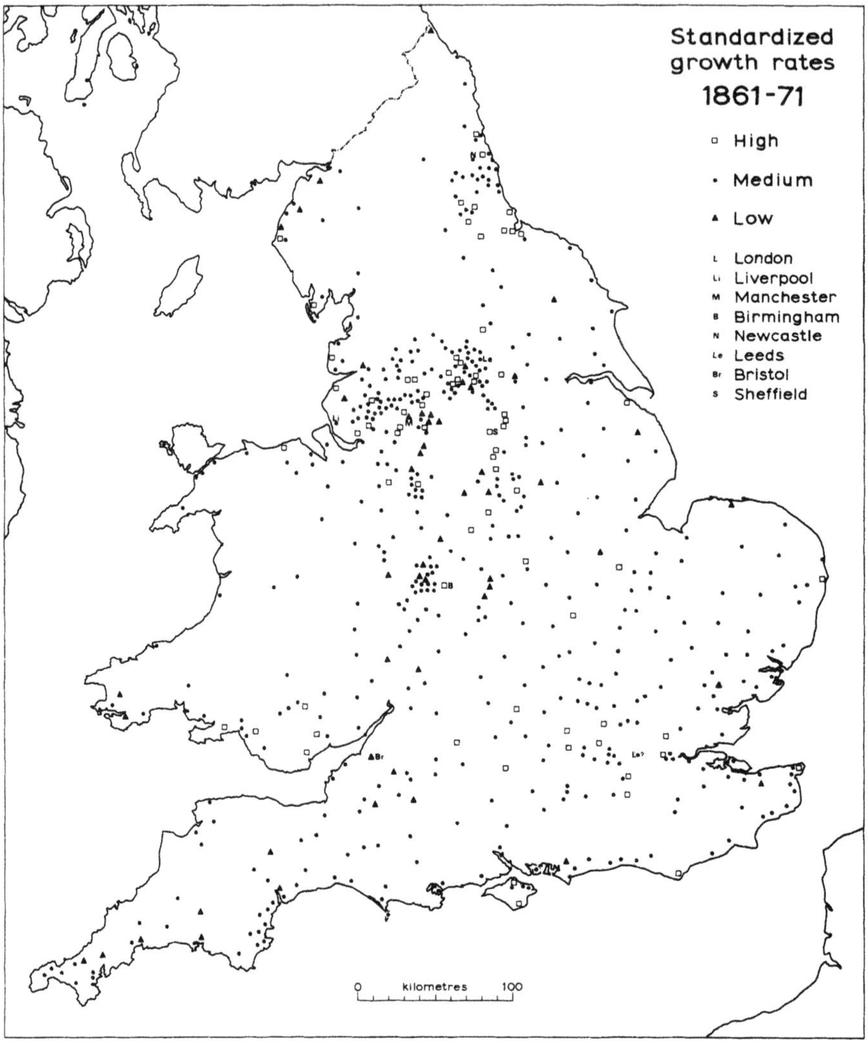

Figure 4.7 The spatial pattern of urban growth in England and Wales, 1861–71. For explanation, see fig. 4.1

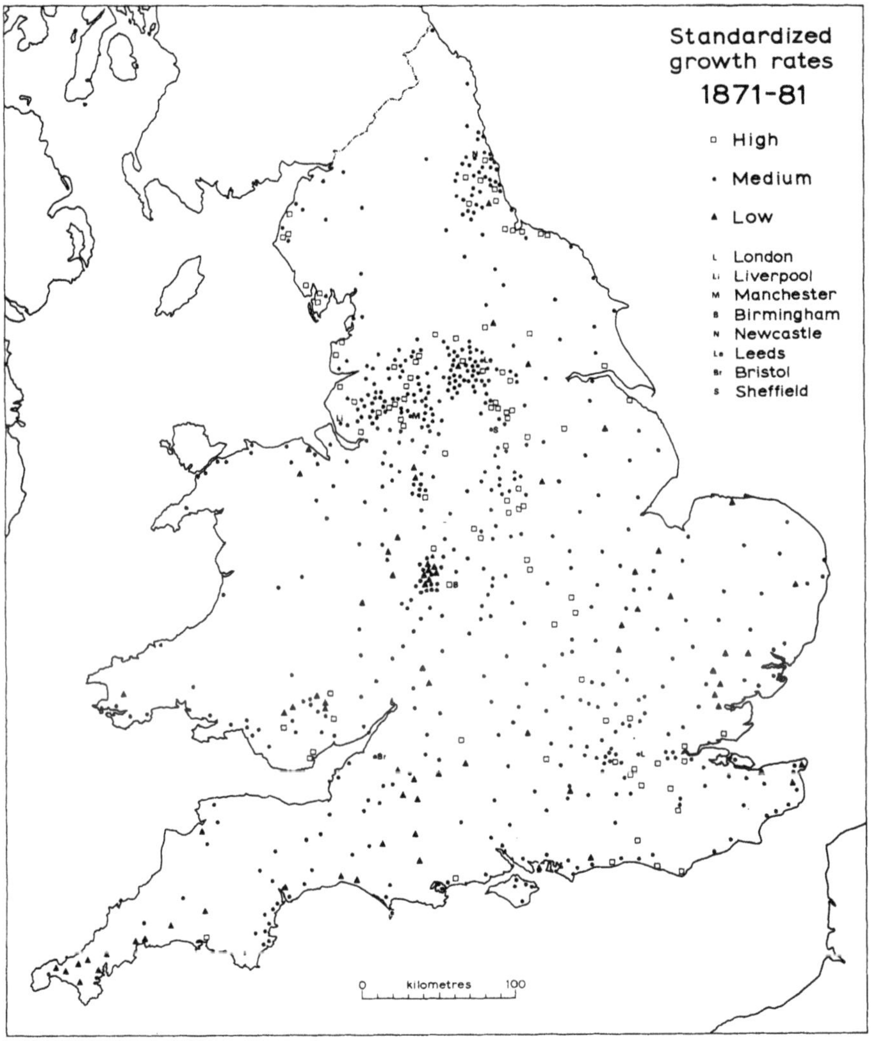

Figure 4.8 The spatial pattern of urban growth in England and Wales, 1871–81. For explanation, see fig. 4.1

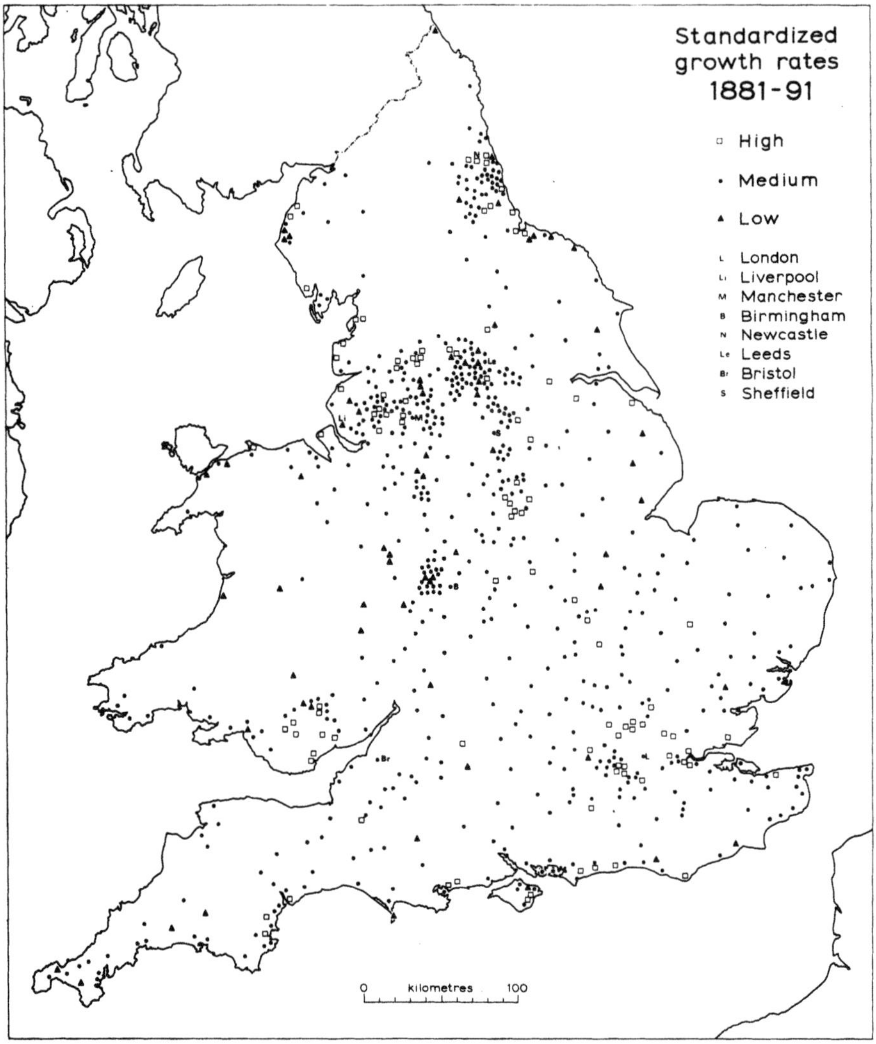

Figure 4.9 The spatial pattern of urban growth in England and Wales, 1881–91. For explanation, see fig. 4.1

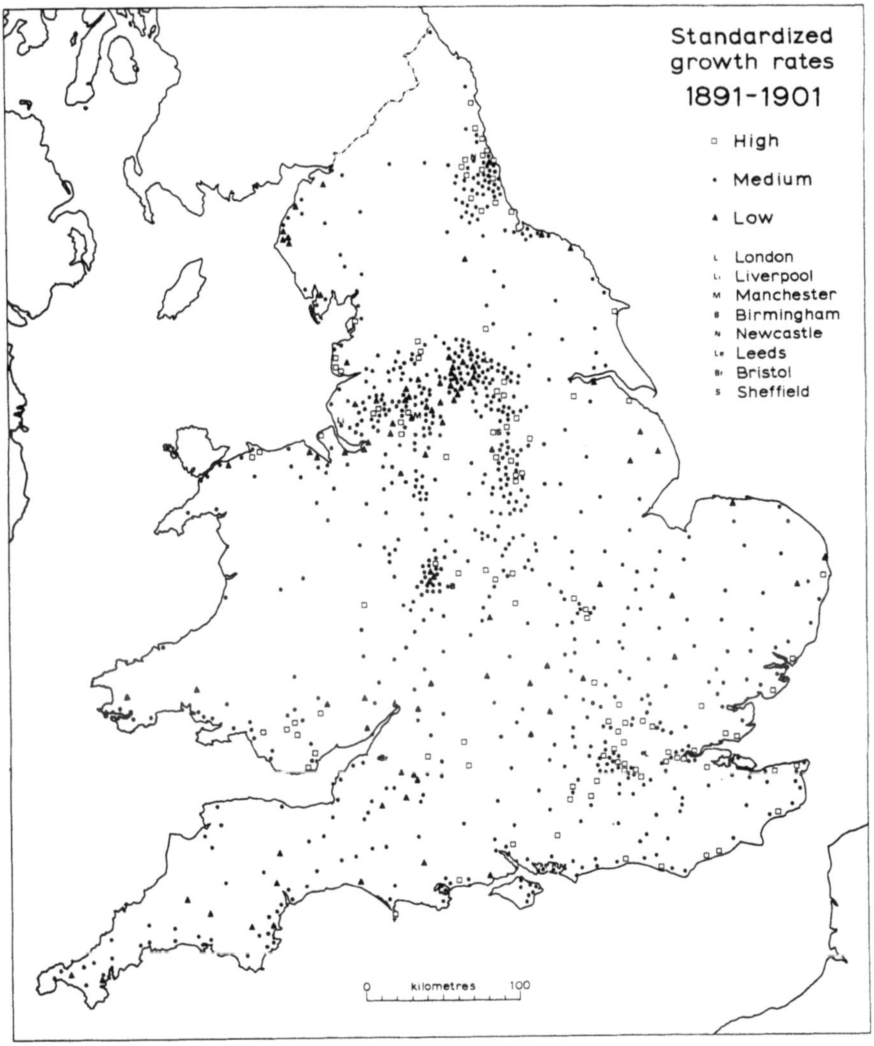

Figure 4.10 The spatial pattern of urban growth in England and Wales, 1891-1901. For explanation, see fig. 4.1

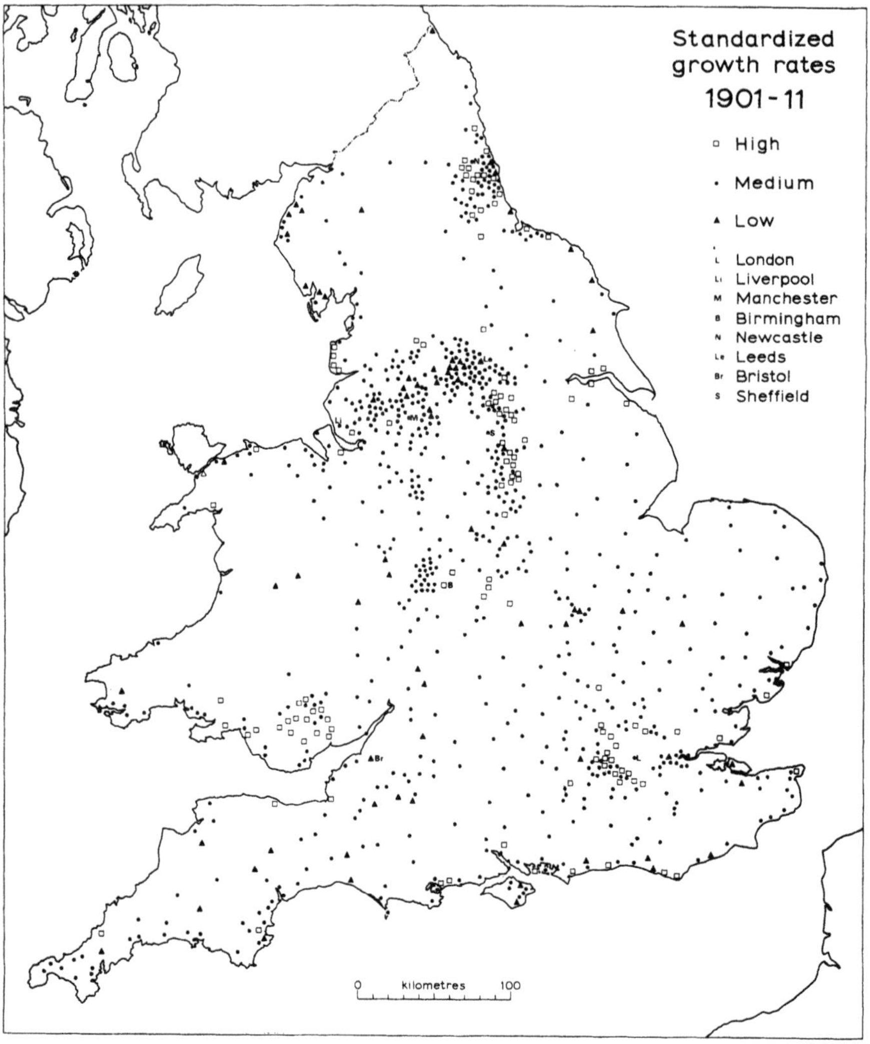

Figure 4.11 The spatial pattern of urban growth in England and Wales, 1901–11. For explanation, see fig. 4.1

development. The earlier decades are dominated by the rapid growth of towns in the textile areas of Lancashire and, at a slightly later date, the woollen area of the West Riding of Yorkshire as well as of the towns of industrial Staffordshire. In the 1820s and 1830s the older textile towns of the West Country and East Anglia decline. After the middle of the century, decline becomes very marked in the more remote agriculturally-based areas, but at the same time newer industrial areas such as Teesside, south Yorkshire and Nottingham-Derby appear. By the final quarter of the century, the metropolitan area begins to show rapid growth while the coalfield areas of South Wales and the North East – for long having had a small scatter of growth towns – both develop strongly as areas of more uniformly high growth. Throughout the whole period, in addition to the differences between the rural and industrial areas and between the metropolitan and non-metropolitan areas, there is a strong undercurrent of growth in resort towns, particularly in coastal resorts. In the early decades, it is the coastal resorts of Sussex which show rapid growth; later, it is the resorts of Devon; and, in the early twentieth century, such striking developments as the growth in the resort towns of north Lancashire are a feature of the two decades following 1891.

It can therefore be said with some confidence that these population changes appear to be useful surrogates of economic growth in general in that the pattern of growth seems to make sense in terms of our knowledge of the industrial and economic history of the period. These qualitative descriptions, however, are very much more the field of the economic or urban historian than the geographer and they would better be amplified by detailed histories of particular industries or particular areas. Can one go beyond this by analyzing the spatial pattern of urban growth? The two hypotheses which have threaded their way as background to this discussion seem to be promising avenues to explore at the somewhat macroscopic scale of our interests. The first is that, particularly in the earlier decades of the century, there should be regions of distinctly high urban growth found in those areas which were undergoing rapid development as the temporal sequence of growth industries caused expansion in distinct regional sub-systems of cities. Second, within any sub-system of cities, one might expect to find a distinctive pattern of spatial growth in which the impetus of growth first occurs in one or a few centres and only subsequently filters out to more distant towns.

REGIONAL GROWTH PATTERNS

The first hypothesis was studied in two ways. First by the use of trend-surface analysis and second by a simpler comparison of the variance of urban growth rates within regions of England and Wales.

Trend-surface analysis is a method of fitting progressively higher-order polynomial equations to a set of x,y,z observations in which the x and y values are co-ordinate data and the z values are observations of some variable, in this case the standardized growth rates.[1] In fitting the surfaces a least-squares solution is used to fit, in two dimensions, first a linear, then a quadratic surface, and so forth. The results, as might be imagined from the complex patterns of the maps of growth rates, were not encouraging. In none of the cases did even the sixth-order surface produce any significant reduction in the sums of squares of the initial observations. This can be seen by looking at some of the results for selected decades. Table 4.1 shows the percentage reduction in the sums of squares for three decades which were chosen as being ones which suggested a relatively simpler pattern than others. Even by the sixth surface, the reduction of the original scatter is only 16, 18 and 10 per cent for the three decades. If one had a

Table 4.1 Trend-surface analysis of standardized urban growth rates

	Percentage reduction in the sum of squares					
	1801–11		*1811–21*		*1831–41*	
Trend-surface	*%*	*Cumulative %*	*%*	*Cumulative %*	*%*	*Cumulative %*
First-order	0·3	0·3	4·6	4·6	0·5	0·5
Second-order	2·4	2·7	0·1	4·7	1·4	1·9
Third-order	6·6	9·3	3·7	8·4	1·1	3·0
Fourth-order	2·1	11·4	2·0	10·4	2·7	5·7
Fifth-order	2·0	13·4	3·0	13·4	1·5	7·2
Sixth-order	2·3	15·7	5·0	18·4	2·3	9·5

Note: Numbers of observations in each decade are as follows: 1801–11 – 255; 1811–21 – 308; 1831–41 – 409.

relatively small number of distinct growth areas in which growth rates of individual towns was high and fairly uniform, one would expect levels of explanation somewhat higher than these figures. It would seem to be the case that, even though certain areas were growing rapidly, the rates of growth of the individual cities which comprise these areas were by no means uniform. Lancashire in the early decades, for example, may have been growing rapidly, but this did not mean that all the towns within, say, the south-east of the county were growing at uniformly high rates, but rather that the area had a higher-than-average proportion of towns which were growing even though at the same time a number of places was growing at well below-average rates. This is exactly the pattern found by Madden (1956b) for nineteenth- and twentieth-century growth rates in American cities of over 10,000 population. He too found that certain areas of rapid growth could be recognized, on the average, but that within each

[1] Trend-surface analysis has been widely used in geomorphology (for example: Chorley and Haggett 1965; Rodda 1970), but has also been applied to the study of the distribution of social phenomena, as for example in the study of urban social structure (Williams 1969).

such region the distribution of growth rates of individual places was scattered throughout every quintile of the overall distribution of growth rates.

The patterns of the trend-surface analysis do, however, appear to confirm our earlier descriptive account of the pattern of growth (fig. 4.12). The macroscopic

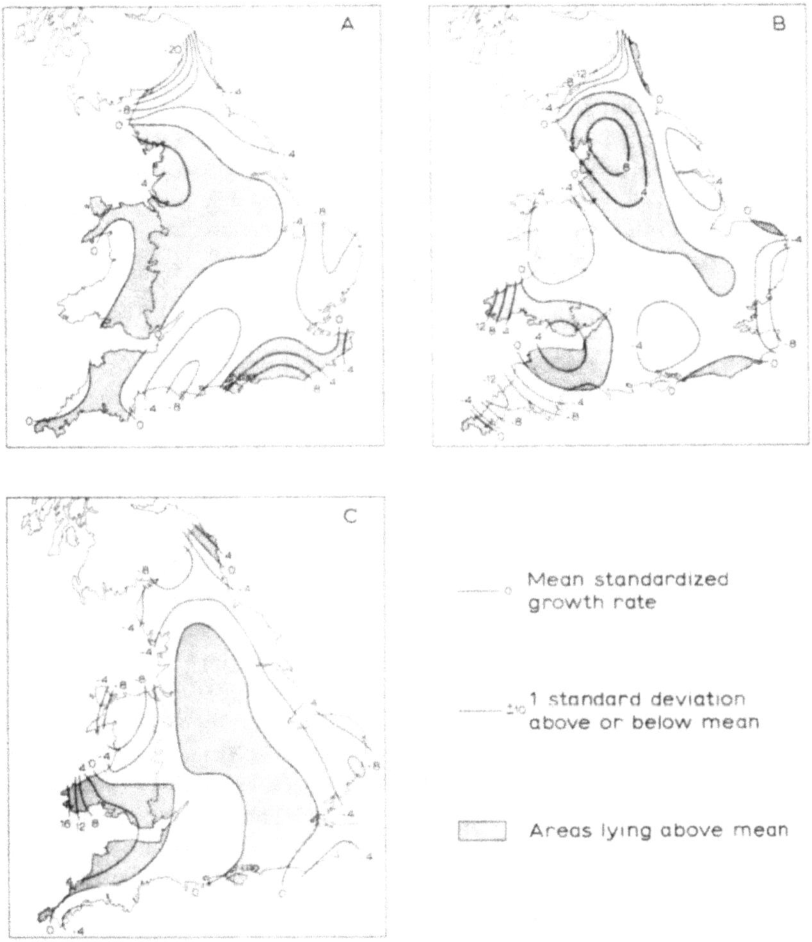

Figure 4.12 Trend surfaces of urban growth rates. Sixth-order surfaces for three selected decades: (A) 1801–11; (B) 1811–21, (C) 1831–41

patterns of the polynomial surfaces show that, in the first decade, the focus of high growth was in the North West with some suggestion of high rates along the south coast; in the second decade the North West again has high rates with markedly lower rates of growth in East Anglia and the South West; and the decade 1831-41 shows a pattern of higher rates roughly in the area of Staffordshire, again accompanied by a belt of high rates along the south coast and with lower rates in the peripheral areas. Interpretation of these surfaces should obviously not be taken very seriously since they evidently 'explain' so small a fraction of the total variance of the original growth rate data. While of some little interest as generalized descriptions of the complex pattern of growth, the trend-surface analysis cannot take us very far in understanding the spatial pattern.

A clearer understanding of the regional complexity of individual urban growth rates can be seen in the second approach to the problem in which averages and standard deviations of growth rates for sets of towns grouped by regions are studied. The thirteen regions which have been selected are simply aggregates of counties made on the basis of their general industrial similarity; they are an attempt to aggregate the whole set of places into a small and relatively homogeneous set of areas. Counties are used as the basic units which are joined purely for the convenience of calculation. It would have been possible to ignore the county groupings and derive more truly homogeneous sets of cities on the basis of a formal regionalization procedure, but, anticipating the findings, this was considered unnecessarily elaborate.

Table 4.2 gives the results of this exercise and, since the growth rates of all places have been included rather than simply isolating those which exceeded plus or minus one standard deviation from the mean, the calculations supplement the spatial distributions portrayed earlier.[1] The full picture of the spatial scatter of growth rates can be seen to be extraordinarily complex. Most of the thirteen regions show a wide scatter of towns in every category of growth rate class. The particular divisions between each class in the table were selected so that each class would include 20 per cent of observations if the growth rates were normally distributed.[2] In fact, in most of the regions and for most of the periods, the percentage distribution of places is little short of an even spread of 20 per cent in each of the five classes of growth rates, suggesting that no area is one with uniformly and significantly high or low growth rates. There are, nevertheless, certain tendencies apparent within each area which give a useful picture of the pattern of urban growth during the nineteenth century. This spatial pattern is clarified if the growth rate data are re-expressed as ratios of the

[1] See above, figs. 4.1 to 4.11.

[2] Thus: the central class runs from minus to plus 0·253, a range which includes 20 per cent of observations under the normal curve; the second and fourth classes run from these limits to figures of minus and plus 0·842; the first and fifth classes are greater than minus and plus 0·842. These figures apply since the standardized growth rate data have been calculated with a mean of 0·0 and a standard deviation of 1·0. It can be seen that the total growth rates of all places at the various decades are not normally distributed, but have a slight positive skew. Had the initial growth rate data been calculated on the basis of simple rather than compound growth, this skew would have been considerably more pronounced. See above, pp. 71-2.

Table 4.2 *Frequency distribution of growth rates by region*

	$\bar{X}$	*s*	% in growth class 1	2	3	4	5	Total no. of towns
NORTH EAST								
1	−0·65	0·92	31·3	50·0	6·3	0	12·5	16
2	−0·16	0·75	27·8	11·1	38·9	5·5	16·7	18
3	−0·35	1·02	28·6	28·6	23·8	4·8	14·3	21
4	−0·39	0·95	31·8	22·7	18·2	18·2	9·1	22
5	0·14	1·38	20·7	20·7	13·8	24·1	20·7	29
6	0·52	1·40	9·1	30·3	6·1	21·2	33·3	33
7	0·41	1·17	9·3	23·3	23·3	16·3	27·9	43
8	0·34	0·96	3·3	27·9	21·3	21·3	26·2	61
9	−0·04	1·05	18·8	31·9	20·3	10·1	18·8	69
10	0·31	1·16	12·8	23·1	21·8	17·9	24·3	78
11	0·43	1·06	9·9	19·8	17·3	21·0	32·1	81
CUMBRIA								
1	−0·42	0·43++	28·6	28·6	42·9	0	0	7
2	−0·06	0·65	12·5	50·0	0	25·0	12·5	8
3	−0·51	0·79	25·0	37·5	12·5	25·0	0	8
4	−0·52	0·78	50·0	25·0	12·5	0	12·5	8
5	−0·50	0·50+	25·0	37·5	25·0	12·5	0	8
6	−0·51	0·41++++	22·2	55·5	11·1	11·1	0	9
7	−0·32	0·91	36·4	36·4	0	18·2	9·1	11
8	0·41	1·23	0	46·2	15·4	7·7	30·8	13
9	−0·16	1·21	28·6	35·7	0	14·3	21·4	14
10	−0·68	0·64+	53·3	26·7	13·3	0	6·7	15
11	−0·89	0·70+	41·2	47·1	11·8	0	0	17
WEST YORKSHIRE								
1	0·14	0·84	9·5	19·0	28·6	23·8	19·0	21
2	0·40	0·86	4·2	8·3	45·8	16·7	25·0	24
3	−0·02	0·85	17·2	27·6	10·3	27·6	17·2	29
4	0·43	1·22	7·0	11·6	27·9	34·9	18·6	43
5	−0·01	0·95	14·6	31·2	27·1	12·5	14·5	48
6	0·17	1·07	7·4	35·2	20·4	14·8	22·2	54
7	0·26	1·03	12·9	22·6	17·7	24·2	22·6	62
8	0·16	0·92	8·9	30·4	17·7	22·8	20·3	79
9	−0·02	0·79++++	13·5	25·8	24·7	24·7	11·2	89
10	−0·15	0·99	22·2	29·3	22·2	16·2	10·1	99
11	−0·01	1·09	13·3	35·2	22·9	11·4	17·1	105
NORTH WEST								
1	0·01	1·22	7·7	5·1	15·4	43·6	28·2	39
2	0·59	1·33	4·4	15·5	24·4	33·3	22·2	45
3	0·22	1·16	17·3	21·2	21·2	21·2	19·2	52
4	0·14	0·97	13·6	22·0	25·4	20·3	18·6	59
5	−0·03	0·89	14·7	26·5	25·0	22·1	11·8	68
6	0·31	1·03	10·6	14·6	32·0	20·0	22·7	75
7	0·33	1·39	12·9	22·3	25·9	14·1	24·7	85
8	0·44	1·06	4·9	20·6	23·5	29·4	21·6	102
9	0·20	1·10	14·0	20·6	27·1	15·9	22·4	107
10	−0·17	1·07	25·6	26·5	24·8	11·1	12·0	117
11	−0·22	0·73++++	13·3	41·7	25·8	13·3	5·8	120

Table 4.2 (continued) *Frequency distribution of growth rates by region*

	$\bar{X}$	s	% in growth class 1	2	3	4	5	Total no. of towns
WEST MIDLANDS								
1	0·11	0·60++	18·5	33·3	14·8	7·4	25·9	27
2	−0·04	0·65++++	9·4	25·0	37·5	18·8	9·4	32
3	0·18	0·72+++	8·1	10·8	40·5	24·3	16·2	37
4	0·46	1·21	10·0	22·5	10·0	22·5	35·0	40
5	0·31	0·84	4·2	18·8	27·1	29·2	20·8	48
6	0·28	1·07	9·6	25·0	23·1	21·2	21·2	52
7	−0·21	0·77+++	16·6	33·3	22·2	20·4	7·4	54
8	−0·13	0·90	16·6	28·3	28·3	20·0	6·6	60
9	−0·32	0·67++++	16·6	38·3	31·6	11·7	1·6	60
10	−0·10	0·79++	14·8	39·3	23·0	9·8	13·1	61
11	−0·20	0·82++	11·5	50·8	18·0	8·2	11·5	61
NORTH MIDLANDS								
1	−0·09	0·80	0	20·0	53·3	6·7	20·0	15
2	−0·19	0·74	15·8	31·6	26·3	15·8	10·5	19
3	0·03	0·62+++	0	38·1	23·8	28·5	9·5	21
4	−0·23	0·56++++	8·7	43·5	26·1	17·4	4·3	23
5	−0·18	0·81	18·5	37·0	29·6	11·1	3·7	27
6	−0·26	0·80	20·7	51·7	6·9	10·3	10·3	29
7	−0·14	0·91	16·6	38·9	22·2	5·6	16·7	36
8	0·39	0·98	6·8	25·0	20·4	15·9	31·8	44
9	0·24	0·80++	1·9	25·0	32·7	17·3	23·1	52
10	0·25	0·79++	3·3	23·0	32·8	21·3	19·7	61
11	0·24	1·04	5·4	36·5	20·2	21·6	16·2	74
EAST								
1	−0·09	0·80	13·0	26·1	30·4	17·4	13·0	23
2	−0·34	0·75++	16·1	32·3	35·5	12·9	3·2	31
3	−0·04	0·62++++	9·1	39·4	21·2	18·2	12·1	33
4	−0·26	0·48++++	10·5	42·1	36·8	7·9	2·6	38
5	0·08	1·05	7·7	28·2	41·0	15·4	7·7	39
6	−0·64	0·65++++	30·2	46·5	16·3	2·3	4·7	43
7	−0·35	0·65++++	15·6	48·9	24·4	6·7	4·4	45
8	−0·38	0·78++	24·4	40·0	13·3	15·5	6·7	45
9	−0·45	0·64++++	29·2	47·9	12·5	4·2	6·2	48
10	−0·28	0·84	24·5	40·8	12·2	8·2	14·3	49
11	−0·15	0·79++	12·0	46·0	26·0	0	16·0	50
SOUTH WEST								
1	−0·05	1·52	24·1	31·0	20·7	6·9	17·2	29
2	0·10	1·24	17·6	29·4	14·7	20·6	17·6	34
3	0·13	1·39	22·0	17·1	24·3	24·3	12·2	41
4	−0·16	0·85	21·2	27·6	25·5	14·9	10·6	47
5	−0·32	0·94	34·6	23·1	23·1	11·5	7·7	52
6	−0·37	0·83+	20·4	46·3	24·1	1·9	7·4	54
7	−0·55	0·60++++	25·4	54·5	9·1	10·9	0	55
8	−0·72	0·63++++	42·9	37·5	12·5	7·1	0	56
9	−0·32	0·68++++	20·7	37·9	29·3	5·2	6·9	58
10	−0·62	0·47++++	35·5	41·9	16·1	6·5	0	62
11	−0·44	0·61++++	27·9	47·1	10·3	7·4	7·4	68

Table 4.2 (continued) *Frequency distribution of growth rates by region*

	$\bar{X}$	s	% in growth class 1	2	3	4	5	Total no. of towns
SOUTH								
1	−0·47	0·61++	26·7	46·7	13·3	6·7	6·7	15
2	−0·11	0·88	20·0	35·0	15·0	10·0	20·0	20
3	−0·33	0·85	19·0	42·9	19·0	4·8	14·3	21
4	−0·03	1·14	20·0	32·0	24·0	12·0	12·0	25
5	0·15	1·27	20·0	24·0	24·0	16·0	16·0	25
6	−0·06	0·92	14·3	42·9	14·3	14·3	14·3	28
7	−0·17	0·77+	12·9	48·4	22·6	6·4	9·6	31
8	−0·53	0·84	37·8	29·7	24·3	2·7	5·4	37
9	−0·02	1·13	23·1	41·0	7·6	10·3	17·9	39
10	−0·03	0·98	20·0	37·8	8·9	13·3	20·0	45
11	−0·08	0·87	17·0	31·9	23·4	17·0	10·6	47
SOUTH MIDLANDS								
1	−0·29	0·42++++	6·2	50·0	25·0	18·8	0	16
2	−0·28	0·57+++	17·6	29·4	23·5	29·4	0	17
3	−0·21	0·64++++	18·5	33·3	25·9	14·8	7·4	27
4	0·01	0·58++++	3·4	44·8	17·2	27·6	6·9	29
5	−0·16	0·84	9·1	51·5	18·2	12·1	9·1	33
6	−0·27	0·67++++	18·2	36·4	24·2	15·2	6·1	33
7	0·04	0·78+	8·6	37·1	17·1	20·0	17·1	35
8	−0·18	0·84	21·6	40·5	10·8	10·8	16·2	37
9	0·02	0·96	12·2	41·4	19·5	14·6	12·2	41
10	−0·09	0·84	27·6	19·1	14·9	23·4	14·9	47
11	−0·33	0·55++++	18·0	46·0	22·0	12·0	2·0	50
SOUTH EAST								
1	0·12	1·01	9·7	16·1	38·7	19·4	16·1	31
2	−0·22	1·04	27·5	37·5	10·0	12·5	12·5	40
3	−0·12	1·06	20·4	36·4	18·2	13·6	11·4	44
4	−0·11	0·83	10·9	37·0	26·1	19·6	6·5	46
5	0·05	0·88	6·0	40·0	22·0	18·0	14·0	50
6	0·16	0·91	4·9	29·5	29·5	19·6	16·4	61
7	0·20	0·82++	5·6	23·9	23·9	32·3	14·1	71
8	0·08	0·98	15·4	29·6	22·0	13·2	19·8	91
9	0·42	1·05	5·8	25·2	13·6	27·2	28·2	103
10	0·53	1·02	4·1	19·7	21·3	24·6	30·3	122
11	0·31	1·12	10·7	20·8	27·7	16·9	23·8	130
WEST								
1	−0·28	0·74	15·3	53·8	0	23·1	7·7	13
2	−0·37	0·68+	25·0	31·3	25·0	18·8	0	16
3	−0·13	0·82	21·1	10·5	42·1	15·8	10·5	19
4	−0·63	0·94	41·7	20·8	25·0	8·3	4·2	24
5	−0·13	1·02	23·3	40·0	10·0	10·0	16·7	30
6	−0·29	0·55++++	8·8	47·1	32·4	5·9	5·9	34
7	−0·42	0·58++++	25·0	38·9	22·2	11·1	2·8	36
8	−0·51	0·58++++	22·5	55·0	10·0	12·5	0	40
9	−0·77	0·67++++	50·0	32·5	15·0	0	2·5	40
10	−0·29	0·79++++	24·4	35·5	17·8	15·5	6·6	45
11	−0·53	0·67++++	31·9	44·7	14·9	4·2	4·2	47

Table 4.2 (continued) *Frequency distribution of growth rates by region*

	$\bar{X}$	*s*	% in growth class *1*	*2*	*3*	*4*	*5*	*Total no. of towns*
SOUTH WALES								
1	0·88	0·82	0	25·0	0	0	75·0	4
2	0·13	0·86	20·0	0	40·0	20·0	20·0	5
3	1·17	1·23	0	12·5	12·5	25·0	50·0	8
4	1·37	1·25	12·5	0	0	25·0	62·5	8
5	0·99	1·20	0	11·1	11·1	44·4	33·3	9
6	0·46	1·16	15·4	7·6	30·7	23·1	23·1	13
7	0·49	1·13	0	35·3	23·5	11·7	29·4	17
8	0·09	1·32	29·6	22·2	7·4	11·1	29·6	27
9	0·64	1·41	13·8	6·9	34·5	10·3	34·5	29
10	0·40	1·14	6·5	25·8	22·6	19·4	25·8	31
11	1·16	1·18	0	10·8	8·1	27·0	54·1	37
DATA FOR ALL REGIONS								
1	0·00	1·00	14·5	27·7	23·0	17·2	17·6	256
2	0·00	1·00	15·9	26·2	25·5	18·8	13·5	309
3	0·00	1·00	16·3	26·9	23·5	19·1	14·1	361
4	0·00	1·00	15·5	28·2	23·5	18·9	13·8	412
5	0·00	1·00	15·2	30·0	24·0	17·6	13·1	466
6	0·00	1·00	13·5	34·2	22·6	14·1	15·6	518
7	0·00	1·00	14·2	33·7	20·8	16·5	14·6	581
8	0·00	1·00	16·6	31·2	18·9	16·9	16·3	692
9	0·00	1·00	16·7	30·3	22·0	14·6	16·3	749
10	0·00	1·00	18·6	28·8	20·7	15·6	16·2	832
11	0·00	1·00	14·3	35·4	20·9	13·4	16·0	887

Notes

1. Standardized (*z*) growth rates are used throughout; the five growth classes of columns 4–8, being as follows:
   1 < −0·842   4 0·253 to 0·842
   2 −0·842 to −0·253   5 > 0·842
   3 −0·253 to +0·253

2. $\bar{X}$ and *s* are, respectively, the means and standard deviations of the standardized rates. Areas whose growth rates are 'bunched' more tightly than those for all regions (i.e. areas showing some homogeneity of urban growth) have standard deviations significantly *lower* than 1·0. These are distinguished (in column 3) for the following confidence levels:
   ++++ 5·0%
   +++ 2·5%
   ++ 1·0%
   + 0·5%

3. For each region the rows refer to decades running from Period 1 (1801–11) to Period 11 (1901–11).

4. Regions are composed of towns from the following groups of counties (abbreviated as in Appendix B):
   North East (DUR, NLD, YNR); Cumbria (CUM, WLD); West Yorkshire (YWR); North West (CHE, LAN); West Midlands (STF, WAR, WOR); North Midlands (DER, LEC, NOT, RUT); East: (CAM, LIN, NFK, SFK, YER); South West (COR, DEV, GLO, SOM); South (DOR, HAM, WIL); South Midlands (BED, BER, BUK, HUN, NTS, OXO); South East (ESS, HRT, KEN, LONDON, MDX, SRY, SSX); West (HFD, SAL, WALES except GLAM, MON); South Wales (GLAM, MON).

appropriate rate for all of the towns within England and Wales as a whole and these figures are shown in table 4.3. A figure greater than 1·0 in any period means that the particular region had a larger percentage of towns falling within the particular class of growth rates than did the whole set of towns for England and Wales during the same period; a figure below 1·0 means that the region had a smaller percentage in that class than did the whole set of towns.

The towns of the North East started off the period with a predominantly low set of growth rates; they had higher growth in the decades from 1841; and by the end of the century the area was clearly one with a predominantly rapid set of growth rates. Cumbria includes very few towns throughout the whole period and its figures are therefore less reliable, but it would appear to have been an area whose towns grew at lower rather than higher rates with the exception of the period 1871-91. West Yorkshire had a rather variable pattern of urban growth, but with a tendency for there to be an over-representation of rapidly growing places throughout most of the period. The North West had a larger-than-expected number of places with rapid growth in all of the decades up to 1891, but thereafter it had more places in the classes of less rapid growth. The West Midlands shows most consistently rapid growth in its cities in the middle decades from 1831-61. The North Midlands appears to have grown most consistently in the latter part of the period, with large proportions of places in the high growth rate classes in the decades from 1871. The East and the South West both show a consistent pattern of predominantly low urban growth; the South West showing this pattern somewhat earlier than the East. The South and the South Midlands show no particularly clear pattern over time. The South East, by contrast, shows a very marked pattern with decline in the period 1811-31, followed by average growth rates in the following decade and increasing proportions of places in the classes of more rapid growth from 1841 onwards. Wales and the Border shows predominant decline during most of the period; while South Wales shows a very distinctive pattern having had high proportions of its towns in the highest growth class for every decade of the period.

These patterns of urban growth in the thirteen regions accord very closely with the fluctuating economic fortunes of their respective regions, but it is quite evident at the same time that there is little suggestion of there being sets of urban regions with homogeneously high or low growth rates. Even in South Wales – the smallest area geographically and the one in which one might expect greatest homogeneity – while there are high proportions of places in the class of the very highest growth rates, these proportions exist alongside significant numbers of places which were found in the classes of lower, or indeed lowest, growth rates.[1] Given this wide scatter of growth rates of towns within regions of England and Wales, it is impossible to talk of regional sets of cities with uniformly high growth rates reflecting the expansion of successive growth industries. The spatial pattern of urban growth is too complex to be teased out by simple spatial or geometric analysis at this aggregate scale.

[1] A scatter which is confirmed by the standard deviations shown in table 4.2.

Table 4.3 *Relative growth frequencies by region*

Period	Ratio of regional % to overall % in each growth class 1	2	3	4	5
NORTH EAST					
1	2·16	1·81	0·27	0·00	0·71
2	1·75	0·42	1·53	0·29	1·24
3	1·76	1·06	1·01	0·25	1·01
4	2·05	0·81	0·77	0·96	0·66
5	1·36	0·69	0·58	1·37	1·58
6	0·67	0·89	0·27	1·50	2·14
7	0·66	0·69	1·12	0·99	1·91
8	0·20	0·89	1·13	1·26	1·61
9	1·13	1·05	0·92	0·69	1·15
10	0·69	0·80	1·05	1·15	1·50
11	0·69	0·56	0·83	1·57	2·01
CUMBRIA					
1	1·97	1·03	1·87	0·00	0·00
2	0·79	1·91	0·00	1·33	0·93
3	1·53	1·25	0·53	1·31	0·00
4	3·23	0·89	0·53	0·00	0·91
5	1·65	1·25	1·04	0·71	0·00
6	1·64	1·62	0·49	0·79	0·00
7	2·56	1·08	0·00	1·10	0·62
8	0·00	1·48	0·82	0·46	1·89
9	1·71	1·18	0·00	0·98	1·31
10	2·87	0·93	0·64	0·00	0·41
11	2·88	1·33	0·57	0·00	0·00
WEST YORKSHIRE					
1	0·66	0·69	1·24	1·38	1·08
2	0·26	0·32	1·80	0·89	1·85
3	1·06	1·03	0·44	1·45	1·22
4	0·45	0·41	1·19	1·85	1·35
5	0·96	1·04	1·13	0·71	1·11
6	0·55	1·03	0·90	1·05	1·42
7	0·91	0·67	0·85	1·47	1·55
8	0·54	0·97	0·94	1·35	1·25
9	0·81	0·85	1·12	1·69	0·69
10	1·19	1·02	1·07	1·04	0·62
11	0·93	0·99	1·10	0·85	1·07
NORTH WEST					
1	0·53	0·18	0·67	2·54	1·60
2	0·28	0·59	0·96	1·77	1·64
3	1·06	0·79	0·90	1·11	1·36
4	0·88	0·78	1·08	1·07	1·35
5	0·97	0·88	1·04	1·26	0·90
6	0·79	0·43	1·42	1·42	1·46
7	0·91	0·66	1·25	0·86	1·69
8	0·30	0·66	1·24	1·74	1·33

Table 4.3 (*continued*) *Relative growth frequencies by region*

Period	*Ratio of regional % to overall % in each growth class* 1	2	3	4	5
9	0·84	0·68	1·23	1·09	1·37
10	1·38	0·92	1·20	0·71	0·74
11	0·93	1·18	1·23	0·99	0·36
WEST MIDLANDS					
1	1·28	1·20	0·64	0·43	1·47
2	0·59	0·95	1·47	1·00	0·70
3	0·50	0·40	1·72	1·27	1·15
4	0·65	0·80	0·43	1·19	2·54
5	0·28	0·63	1·13	1·66	1·59
6	0·71	0·73	1·02	1·50	1·36
7	1·17	0·99	1·07	1·24	0·51
8	1·00	0·91	1·50	1·18	0·41
9	0·99	1·26	1·44	0·80	0·10
10	0·80	1·37	1·11	0·63	0·81
11	0·80	1·44	0·86	0·61	0·72
NORTH MIDLANDS					
1	0·00	0·72	2·32	0·39	1·14
2	0·99	1·21	1·03	0·84	0·78
3	0·00	1·42	1·01	1·49	0·67
4	0·56	1·54	1·11	0·92	0·31
5	1·22	1·23	1·23	0·63	0·28
6	1·53	1·51	0·31	0·73	0·66
7	1·17	1·15	1·07	0·34	1·14
8	0·41	0·80	1·08	0·94	1·95
9	0·11	0·83	1·49	1·19	1·42
10	0·18	0·80	1·59	1·37	1·22
11	0·38	1·03	0·97	1·61	1·01
EAST					
1	0·90	0·94	1·32	1·01	0·74
2	1·01	1·23	1·39	0·69	0·24
3	0·56	1·47	0·90	0·95	0·86
4	0·68	1·49	1·57	0·42	0·19
5	0·51	0·94	1·71	0·88	0·59
6	2·24	1·36	0·72	0·16	0·30
7	1·10	1·45	1·17	0·41	0·30
8	1·47	1·28	0·70	0·92	0·41
9	1·75	1·58	0·57	0·29	0·38
10	1·32	1·42	0·59	0·53	0·88
11	0·84	1·30	1·24	0·00	1·00
SOUTH WEST					
1	1·66	1·12	0·90	0·40	0·98
2	1·11	1·12	0·58	1·10	1·30
3	1·35	0·64	1·03	1·27	0·87
4	1·37	0·98	1·09	0·79	0·77
5	2·28	0·77	0·96	0·65	0·59

Table 4.3 (*continued*) *Relative growth frequencies by region*

	Ratio of regional % to overall % in each growth class				
Period	*1*	*2*	*3*	*4*	*5*
6	1·51	1·35	1·07	0·14	0·47
7	1·79	1·62	0·44	0·66	0·00
8	2·58	1·20	0·66	0·42	0·00
9	1·24	1·25	1·33	0·36	0·42
10	1·91	1·46	0·78	0·42	0·00
11	1·95	1·33	0·49	0·55	0·46
SOUTH					
1	1·84	1·69	0·58	0·39	0·38
2	1·26	1·34	0·59	0·53	1·48
3	1·17	1·60	0·81	0·25	1·01
4	1·29	1·14	1·02	0·64	0·87
5	1·32	0·80	1·00	0·91	1·22
6	1·06	1·25	0·63	1·01	0·92
7	0·91	1·44	1·09	0·39	0·66
8	2·28	0·95	1·29	0·16	0·33
9	1·38	1·35	0·35	0·71	1·10
10	1·08	1·31	0·43	0·85	1·24
11	1·19	0·90	1·12	1·27	0·66
SOUTH MIDLANDS					
1	0·43	1·81	1·09	1·09	0·00
2	1·11	1·12	0·92	1·56	0·00
3	1·14	1·24	1·10	0·78	0·53
4	0·22	1·59	0·73	1·46	0·50
5	0·60	1·72	0·76	0·69	0·70
6	1·35	1·06	1·07	1·08	0·39
7	0·61	1·10	0·82	1·21	1·17
8	1·30	1·30	0·57	0·64	0·99
9	0·73	1·37	0·89	1·00	0·75
10	1·48	0·66	0·72	1·50	0·92
11	1·26	1·30	1·05	0·90	0·13
SOUTH EAST					
1	0·67	0·58	1·68	1·13	0·92
2	1·73	1·43	0·39	0·67	0·93
3	1·25	1·35	0·77	0·71	0·81
4	0·70	1·31	1·11	1·04	0·47
5	0·40	1·33	0·92	1·02	1·07
6	0·36	0·86	1·31	1·39	1·05
7	0·39	0·71	1·15	1·96	0·97
8	0·93	0·95	1·16	0·78	1·22
9	0·35	0·83	0·62	1·86	1·73
10	0·22	0·68	1·03	1·58	1·87
11	0·75	0·59	1·33	1·26	1·49
WEST					
1	1·06	1·94	0·00	1·34	0·44
2	1·57	1·20	0·98	1·00	0·00
3	1·29	0·39	1·79	0·83	0·75

Table 4.3 (*continued*) *Relative growth frequencies by region*

	Ratio of regional % to overall % in each growth class				
Period	*1*	*2*	*3*	*4*	*5*
4	2·69	0·74	1·06	0·44	0·30
5	1·53	1·33	0·42	0·57	1·28
6	0·65	1·38	1·43	0·42	0·38
7	1·76	1·15	1·07	0·67	0·19
8	1·36	1·76	0·53	0·74	0·00
9	2·99	1·07	0·68	0·00	0·15
10	1·31	1·23	0·86	0·99	0·41
11	2·23	1·26	0·71	0·31	0·26
SOUTH WALES					
1	0·00	0·90	0·00	0·00	4·26
2	1·26	0·00	1·57	1·06	1·48
3	0·00	0·47	0·53	1·31	3·55
4	0·81	0·00	0·00	1·32	4·53
5	0·00	0·37	0·46	2·52	2·54
6	1·14	0·22	1·36	1·64	1·48
7	0·00	1·05	1·13	0·71	2·01
8	1·78	0·71	0·39	0·66	1·82
9	0·83	0·23	1·57	0·71	2·12
10	0·35	0·90	1·09	1·24	1·59
11	0·00	0·31	0·39	2·02	3·38

Note: The growth classes, the periods and the composition of regions are as given in table 4.2

SPATIAL PATTERNS: GROWTH RATES AND DISTANCE

The second hypothesis suggests that growth impulses in a set of cities will be filtered out from certain cities to others so that a temporal lag will occur between their respective growths or that growth rates will decline at increasing distances from the growth centre.[1] Once again, such an hypothesis derives from the notion that the interconnections between towns form urban systems and sub-systems. The most impressive studies which have attempted to disentangle patterns of leads and lags in sets of cities have used rates of unemployment as indicators of economic health. King, Casetti, and Jeffrey (1969), for example, have postulated a model in which the activity levels in a given city are the effect, first, of levels of activity in other cities and, second, of the effects of national factors whose consequences are delayed by various time lags, the duration of which depends on the linkages of the given city. Looking at unemployment data

[1] These are, of course, different notions: the first suggesting a temporal and the second a spatial pattern. However, both are derived from the hypothesis that the influence of towns upon each other are channelled through a network of linkages which help define the system of cities. In the ensuing 'test' of the hypothesis it is assumed that these linkages are spatially determined. With better data, the more realistic assumption that the linkages follow the hierarchy of urban size could be studied in the way in which it has been examined for present-day Swedish towns by Wärneryd (1968).

for 1960-64 for midwestern cities of the United States, their analysis isolates three sub-systems of cities centred on Pittsburgh, Detroit and Indianapolis; the latter two lagging behind the first. Other cities within the area show less interaction since their linkages are less with the regional than with the national urban system.[1] At a finer scale, Bassett and Haggett (1971) have studied the pattern of leads and lags in towns in south-west England using lag correlation and cross-spectral methods to analyze the time series of monthly unemployment figures. Their results suggest a complex pattern of leads and lags amongst the set of places with the upturns and downswings of such places as Swindon, Stroud and Bath leading those of Bristol by periods of from six to one months and Gloucester, on the other hand, lagging by some two months. Analyses of this sort have suggested most interesting methods of unravelling the economic interlinkages between towns, but they are extremely demanding in terms of data requirements (Granger 1964). In studying the growth of cities of the nineteenth century, one set of data which would seem to provide observations of a sufficiently fine temporal mesh are the annual statistics on house building collected by Weber for a large number of individual towns in Britain. Parry Lewis (1965), in looking at these data, himself suggests the potential value of studying inter- and intra-regional lags in the movements of the individual house-building series. In the building boom of the 1890s, for example, the great majority of areas followed the national trend with a downswing occurring in the early part of the 1900s. South Wales, however, showed a markedly different regional pattern with house building continuing at high rates into the late 1900s.[2] Even such data on house building, however, provide observations only at annual intervals and, since the connections between rates of building and population increase are by no means in happy accord and since house building is in competition with other economic activities for many of the same factors of production so that many of its swings move inversely to indicators of more general economic activity, it would seem that a study of house-building activity rates would not reveal much about the demographic expansion of urban areas.[3]

[1] Chicago is one of these latter, nationally-orientated, cities. Such a distinction accords neatly with the national and regional dichotomies of the American urban system which are noted above, pp. 91–2.

[2] This difference might be seen as a consequence of the continuing high rates of migration into South Wales. Maps of net migration flows (Friedlander and Roshier 1966) show that the dramatic reversal of migration in South Wales occurred after the decade 1901-11; up to 1911 immigration into the area was very heavy, but after 1911 outmigration to the South East began in large numbers. See especially B. Thomas (1930).

[3] The relationship between house building and population pressure (as measured by the number of empty houses) has been studied by Saul (1962a, p. 129) who suggests that 'though building was normally influenced by the level of empties obtaining, at the height of a boom there may be a considerable lag before building was reduced in the face of the high empties and the effects could also be much delayed before the upswing got under way.' See also the similar conclusion reached by Cairncross (1953, pp. 27-33). The most generally agreed opinion is that building activity appears to be less a direct indicator of economic expansion than of migration. Cairncross (1953, p. 25) called the building cycle 'little more than a migration cycle in disguise'.

An approach to the analysis of leads and lags in rates of urban growth which provides a more spatial focus and is less taxing in its data requirements is provided in the study by Casetti and Semple (1968) of urban growth in small towns in southern Ontario. They found that growth rates tend to decline in magnitude with increasing distance from a limited number of growth centres in the area: three centres being found which formed an arc of growth around the head of Lake Erie from which, at increasing distances, the growth rates of other towns fell; two further centres were found to the north-west of this from which, at increasing distances, growth rates rose. Impressive amounts of explanation were provided by the isolation of spatial growth trends from these five points. Distance from the first growth point alone explained almost 70 per cent of the total variance in the growth rates during the decade 1950-60; and almost 95 per cent of the variance was explained by the joint 'effects' of all five points.

The technique suggested by Casetti and Semple was applied to one of the decades in the sequence of urban growth in England and Wales. The method used involves superimposing over the whole area a grid of x,y co-ordinates; from each intersection of the grid, linear equations are then fitted expressing the growth rates of all towns as some function of distance from the intersection; the intersection with the best fit is then taken as the point from which growth either declines or increases; the residuals from this regression equation are then used as observations for the fitting of subsequent equations, and thus the choice of a second, a third and nth points from which growth either decreases or increases. One therefore picks out points which successively explain the largest amount of the observed growth (or residuals) in terms of distance decay or distance accretion. For an area as large as England and Wales, it would be unreasonable to anticipate that growth might be related to distance from a single point, but some form of transformation of the distances would obviously enable one to steepen the fall-off in the distance effect and so isolate regional growth centres.

The results of applying this technique to the growth rate were not encouraging. Table 4.4 shows the results for the decade 1801-11 and it can be seen that the amounts of explanation which are afforded are extremely small. This might be expected from all that has so far been said of the complex spatial pattern of growth during the nineteenth century and the results are yet again confirmation of the absence of any simple geometrical pattern of growth rates at this macroscopic scale of analysis. The points which are selected are nevertheless of some small interest since, despite their low reliability, they do make some sense in terms of the details of the growth rates of the chosen decade. In all cases, the various permutations of the growth rates – either the raw growth rate data or the standardized rates – and the distance transformations – either using a reciprocal or a logarithmic transformation – make little difference to the patterns which emerge. The first growth point which is isolated falls in Lancashire with a tendency for growth to decrease at increasing distances away from it. The second point, isolated both by the first and second tests, falls within an area on the coast of Kent, again from which point growth rates tend to fall. The two other points which are isolated fall in the vicinity of Whitby in Yorkshire and within the west of Wales, from one of which growth rates tend to rise. It is interesting to note that the points in Lancashire and in Kent accord with what has already been seen of the spatial pattern of growth rates in this decade, but, given the amount of explanation provided by these four points, it is evident that one would need to isolate a very large number of points before being able to account for any large fraction of the total variance. The conclusion must be that, at this spatial scale, the pattern of urban growth is too complex for such techniques to have any great analytic value. Only at a much more detailed

Table 4.4 *Growth centres, 1801–11*

Location (national grid references)	*Correlation* (r)	*Cumulative* (r^2)	*Distance effect*
(*a*) *Reciprocal of distances and raw growth rates*			
1st centre 371.416 (near Bury, Lancs.)	0·22	4·7	decrease
2nd centre 638.155 (near Dover, Kent)	0·17	7·6	decrease
3rd centre 260.251 (central Cardiganshire)	0·12	8·9	decrease
(*b*) *Reciprocal of distances and standardized growth rates*			
1st centre 371.413 (near Bury, Lancs.)	0·23	5·4	decrease
2nd centre 638.155 (near Dover, Kent)	0·16	8·0	decrease
(*c*) *Logarithm of distances and standardized growth rates*			
1st centre 391.407 (near Oldham, Lancs.)	−0·23	5·1	decrease
2nd centre 488.513 (near Whitby, Yorks.)	0·16	7·5	increase

Notes

1. For explanation of method, see text.
2. The final column shows whether urban growth rates tend to increase or decrease at increasing distances from the selected 'growth' centre.

level of study – looking at the distribution of growth rates both at a finer temporal scale and within much smaller areas of the country – might one unearth any tendency for spatial regularity.

CONCLUSION

The various analyses of the macroscopic spatial pattern of urban growth have produced uniformly negative results. Even though the generalized patterns do accord with knowledge of the regional economic growth of England and Wales, the details of the spatial patterns can only be interpreted in terms of the particular factor endowments and historical events which underlay the growth of certain towns and the decay of others. There is little in the spatial pattern, at the scale at which it has been studied, which could be interpreted better by a geographer than by the descriptive expertise of the historian. At the national scale, it appears not to be the case that there were regions (or *spatial* sub-systems) which underwent uniformly rapid growth or decline during the nineteenth century. Rather, particular areas showed rapid average rates of urban growth because they had a higher-than-average number of rapidly growing towns, but at the same time they also had significant numbers which grew at rates similar to or below the national average. Nor, again, at the scale at which we have looked, can it be said that there are key growth points at greater distances from which the growth rates of the majority of other towns decline. It may well be that such patterns could be found at a smaller scale of study or with population data for periods less widely spaced in time than a decade. The latter information cannot be derived for any large number of places and the former approach lies outside the national frame of reference adopted in the present study. All that can be said in general terms is that the spatial pattern of nineteenth-century urban growth shows most dramatically the very marked shift from the early boom of the northern textile and metal-working areas and the much later growth of the towns within the South East. Before one could argue that this general pattern is suggestive of the early development of distinct sub-systems and the later development of a more unitary system focused on London, one would need both more sensitive indicators of the existence of systems, as well as a good deal of non-Euclidian spatial ingenuity.

We must, therefore, return to the two unresolved problems of urban growth which were outlined in the last chapter in an attempt to explain the scatter of urban growth rates and the particular decline of the average growth rates of large places in the final third of our period.

PART 3

Approaches to explanation

CHAPTER 5

Technical innovations and the scatter of urban growth

He that will not apply new remedies must expect new evils.

FRANCIS BACON

The first aspect of urban growth to which we shall turn is the decreasing scatter of growth rates which are found at increasing city sizes. This is the more interesting because it is the more general of our two findings from the data of nineteenth-century cities. There are many possible lines of explanation which could be explored. First a purely statistical approach could be adopted and it could be argued that large units might be expected to have small variances simply on the basis of elementary statistical theory. If growth is the result of the addition of employment opportunities, a slight fluctuation in the number of factories added to a town of 1,000,000 will make less proportionate difference to its growth than it would in the case of a town of 10,000. The arrival of one new firm would barely ruffle the smooth growth curve of Birmingham, but could cause a marked upswing in the curve of Bewdley. Second, and as amplification of this first statistical explanation, it could be argued that towns are a collection of some more fundamental 'unit' at a lower level of scale[1] so that a very small town might be seen as consisting of the summation of, say, 100 such 'units', a medium sized town would be composed of 1,000 units and a large town of 100,000 units. Then, while the variance in the growth of each of these units would be the same, the greater the number of units the less would be the overall variance on the assumption that, the distribution of probabilities of growth of each unit being normal, the addition of increasing numbers of these units would produce an overall growth closer and closer to the average.[2] Such an approach, however, would suffer from the fact that it is difficult to identify the basic units so that the explanation is singularly unrevealing of the nature of urban growth. Whereas in the case of the growth of firms, one can distinguish the respective units of divisions and branches of an overall firm which could be considered to

1 The problems of definition which would be involved here hark back to our earlier discussion of defining objects at successive levels of scale.

2 An explanation which has been considered in the parallel case of business-firm growth rates (Singh and Whittington 1968, p. 93, footnote 2).

form the basic units, the constituent parts of an urban economy are both more numerous and less easily identifiable.

An alternative approach might be to argue that large cities are the organizing poles of a regional economy and that much of their economic vitality derives from the commercial activities involved in the servicing and marketing functions which they offer to the more specialized subservient industrial towns within their hinterland. Thus, while the upswings and downswings affecting particular industries would have marked effect on a town whose economy was closely tied to that particular industry, the large regional centre would continue to flourish on the basis of the many activities which it performed for other towns not dependent on the declining industry. While it is certainly the case that many of the large towns had a greater variety of industrial activities within them than did smaller places – Birmingham's welter of trades being the prime, although slightly unusual, instance – the absence of detailed occupational or industrial data for towns in the nineteenth century means that it would be impossible to estimate with any accuracy the proportions of industrial and commercial activities within individual towns.[1]

Instead, the line of argument which will be pursued here turns to the connection between urban growth and the spread of innovations which can be argued to underlie the economic vitality of towns. It will be argued that the characteristics of entrepreneurial diffusion can produce patterns of growth rates which conform to our empirical finding that the variance of growth decreases with the increasing size of towns.

In the aftermath of Schumpeterian writings, there has been a widely-accepted belief in the importance of technological innovations rather than capital inputs as a factor in urban growth. In the city of today, growth is as much a product of tertiary activities as of manufacturing and, with the growth in white-collar occupations, the largest cities are now characterized by very extensive office development and the growth of service rather than manufacturing industry. For the nineteenth century, this was much less the case. Although London had for long been the most important commercial centre in the country, it too was a very substantial centre of industry – the largest in the country in absolute terms – and industrial activity together with its associated activities, particularly transportation, provided work for very considerable numbers of the employed

[1] While occupational data are given in the earlier censuses, the early classifications are far from satisfactory (Booth 1886). Only in 1911 – and more fully in 1921 – was a separate distinction drawn between occupational and industrial classifications. For periods before this, see the heroic attempts of Welton (for example: 1868-9; 1869; and 1897-8) to work towards the current notion of primary, secondary and tertiary activities. For an attempt to derive occupational data for a single town during the second half of the nineteenth century, see Bellamy (1952). The problems and potentials of using such census data are well discussed by Armstrong (1972).

population.[1] For the majority of towns it was industry which provided the bulk of employment opportunities of nineteenth-century towns. And the advent of new types of industries and the growth of existing ones depended very much upon the development and application of new industrial expertise, which either improved existing practices and so increased production relative to inputs of capital, labour and materials, or else introduced entirely new types of economic activity. Both of these types of expansion in industrial activity within towns would tend to be self-generating in that they would widen markets and set the stage for further expansion in production and they would also encourage the development of new activity in a host of related types of manufacturing and servicing activity which were linked to the original industry by 'forward' or 'backward' linkages. The application of the ideas of Bessemer and Siemens, for example, made possible the production of vast quantities of relatively cheap steel in the 1860s and, more especially, the 1870s. But this in turn had consequent effects on activities which handled the ores and fuels, on those which used steel both in heavy construction, as in shipbuilding, and in the lighter finishing trades, and all of these activities in turn had effects on the growth of the multitude of services called into play to provide housing, urban facilities, transportation, capital and the host of infrastructural and service requirements called for by the additional labour so produced.

Ramifications such as these may have had effect either within a single town or more generally, but the nature of the process is clear; there is a form of positive feedback effect in which growth calls forth additional growth in a cumulative and circular fashion. Models of this process seem themselves to have been subject to the same positive feedback effects since there is an enormous literature on cumulative causation both at regional and sub-regional scales. For the growth of cities, a relevant suggestion is that proposed by Pred (1966) which is illustrated in fig. 5.1. He envisages two feedback loops which are associated with an initial growth impetus produced by new or enlarged industrial activity within a city during a period of manufacturing predominance. The first of these loops is traced through an initial multiplier effect which is the product of the growth of the manufacturing function itself plus the growth of manufacturing activities with which it is linked. Associated with this is a secondary multiplier effect caused by the concomitant expansion in construction and tertiary activities. This first loop leads to the expansion of the local or regional market on which, Pred claims, the manufacturing industry is dependent; so that, with expansion, new threshold levels are reached and growth in the industry and its labour force continues to cumulate. The second loop is associated with the

1 Hall (1962, pp. 21 and 183-5) calculates that, in 1861 the percentages employed in primary, manufacturing and service activities were respectively, 24·0, 33·2 and 39·2 for England and Wales and 3·0, 31·7 and 61·0 for London. One-quarter of the service activities, however, is comprised of building and transport employment.

argument that growth in production increases the likelihood of further new inventions and innovations within the local economy and so leads to yet further growth which will be produced by them. Such a notion of positive and cumulative causation in the growth of local manufacturing economies seems widely to be accepted and there is certainly evidence that technological innovations generate fresh industrial activity and in turn are more likely in the context of a vibrant than a stagnant economy.[1]

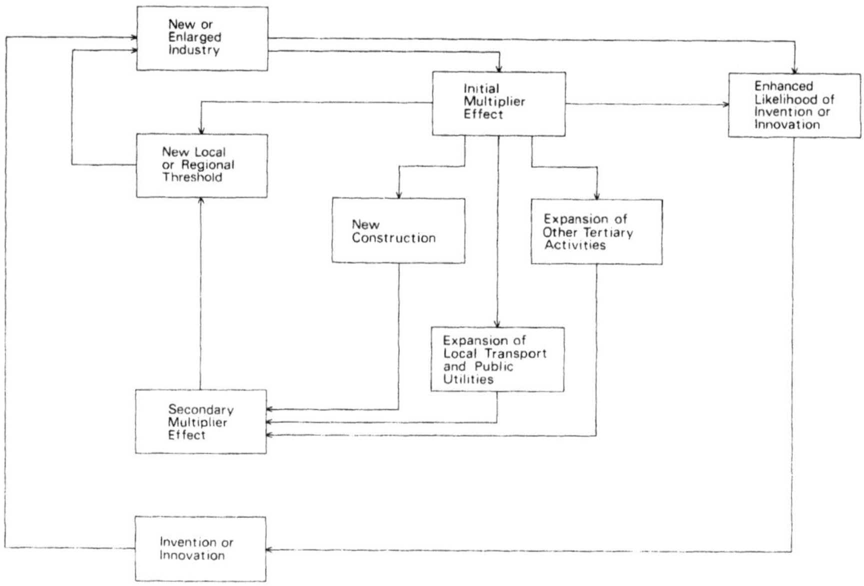

Figure 5.1 The process of urban growth: circular and cumulative causation. *Source:* Pred (1966)

It is this concept which must partly be held responsible for the considerable interest which has been engendered in the manner by which innovations diffuse throughout an area or an economy. The most voluminous of this literature deals with the temporal and structural aspects of innovation diffusion rather than its spatial characteristics. Rogers' (1962; see also Rogers and Shoemaker 1971) early summary of literature draws most strongly on work within agriculture and the social sciences and is still an invaluable compilation of evidence of the

[1] Schumpeter (1939) argued that innovations appear in 'clusters' over time as bottlenecks in related industries spurred the development and application of new techniques within industries which were linked in their productive processes. There is some disagreement amongst economists as to whether innovations are developed more in periods of slack or of full activity (see for example, Mansfield 1969, pp. 106-7).

existence of a logistic curve of acceptances of an innovation over time and also of the relationship between the characteristics of the adopters and the date at which they adopt – whether they are 'leaders', 'early-' or 'late-majority', or 'laggards' in the temporal process. There is in addition a flourishing literature in epidemiology (for example, Bailey 1957) and in the spread of technology (see, for example, the summary in Mansfield 1969). It is, however, the spatial aspects of diffusion with which our main concern will lie and, although late in developing, there is now an accumulating body of empirical and theoretical literature on which to draw (Brown 1968). Hägerstrand was the first to contribute a distinctly spatial approach to the formal analysis of the diffusion of innovations and his *Spatial diffusion of innovations* has already become one of the few 'classics' of modern geography because of its careful unfolding of a spatial argument and its incorporation of more general spatial theory.[1] The focus in this and much subsequent geographical study of diffusion has been on the process of contagious diffusions. The acceptance of TB controls amongst farmers, the spread of pastoral subsidies, of foot and mouth epidemics, or of tractors are all processes which can be seen in terms of a progressively advancing frontier of diffusion. In such studies the logistic curve of the temporal process of adoption can be directly applied to the spatial case. The logistic expression is given by the following:

$$P = \frac{U}{1 + e^{a - bT}}; \qquad [5.1]$$

where P is the percentage of adopters; U is some upper saturation limit; T is time; and a and b are parameters. With spatially-continuous diffusion, this equation can readily be made to apply to the spatial rather than the temporal case by expressing the a and b parameters as some function of distance from the place of first adoption – as in Casetti and Semple's (1969) study of the spread of tractors in a rural area. Alternatively, as in Hägerstrand's case, the spatial spread can be simulated in a Monte-Carlo fashion by which the probability of adoption is made a function of distance from individuals who have already adopted.

The logic of such spatially-continuous diffusion models is that, first, there is a relatively uniform distribution of population 'at risk' and, second, that the likelihood of adoption is a function of distance because it is determined by face-to-face contacts through the network of private communications which link individuals together. Such assumptions may meet the case of many agricultural innovations where the variance in the size of farms may not be large, where their distribution is relatively uniform and where, even after the widespread development of the mass media of communications, the decision to

1 Although see the criticism of Hägerstrand's interpretation of patterns of 'neighbourhood' spread in Cliff (1968) and Cliff and Ord (1970, pp. 285-9).

adopt (as distinct from the fact of *knowing* about the innovation)[1] may still depend on face-to-face contacts. The case of adoption of innovations by towns, however, seems not to meet such requirements. As adopting 'units', towns are both discretely located in space and there are much more marked differences in their sizes than is the case for either farms or private households. Furthermore, many of the relevant innovations are of a different sort from those which have been studied in the contiguous diffusion of agricultural innovations. It is useful, here, to differentiate between what Pederson (1970) calls entrepreneurial and household innovations: the first being those which are adopted by organizations or by composite decision-making bodies; and the second those adopted by individual households. The difference is illustrated, for example, by the spread of television stations on the one hand and the spread of individual television sets on the other. The latter is obviously contingent upon the prior spread of the former, but there are more fundamental differences than this. The entrepreneurial innovation is one which will usually involve a larger capital outlay and a greater element of risk and is more likely to be based on a somewhat more 'rational' economic assessment of the appropriate market potential of the innovation. They also differ in that the entrepreneurial innovation is essentially competitive; whereas the fact that a household adopts a television set in no way precludes the adoption of sets by other households – indeed the essence of the diffusion argument is that it is likely to encourage rather than discourage such further adoption – once a firm or company establishes a television station in a city the probability of further companies setting up stations is more likely to be reduced than enhanced. Only when the market potential for a particular innovation is sufficiently great to support more than a single entrepreneurial innovation is one likely to find more than a single adoption in a given town. For commercial entrepreneurial innovations, the size of the potential market is therefore a consideration of the first importance and the size of the adopting unit – the town – is therefore critical. This adds a new element to the emphasis on space and distance. In looking at the diffusion of innovations in sets of towns it would appear that both size and distance need to be taken into consideration.

The importance of size of town in the spread of urban innovations has long been noted (Bowers 1937; McVoy 1940; Crain 1966), and indeed, Hägerstrand himself recognized the need to incorporate the influence of the size hierarchy into any study of diffusion within a set of towns. He distinguishes between a 'neighbourhood' effect on the one hand and an 'hierarchical' effect on the other: the former being the process of contiguous spread which geographers have studied in some detail; the latter being a diffusion from larger to smaller places.

[1] A distinction well illustrated by Ryan and Gross (1943). The stages at which mass media are of great importance are the very early ones of 'awareness' and 'interest'. In the later stages of 'evaluation', 'trial' and eventual 'adoption' the role of personal contact through friends and neighbours becomes paramount. See, for example, Mansfield (1969, p. 129).

He argues (Hägerstrand 1966, p. 40) that the 'frontier of urban diffusion' will be found in the successive ranks of the hierarchy of towns:

> The point of introduction in a new country is its primate city; sometimes some other metropolis. Then centres next in rank follow. Soon, however, this order is broken up and replaced by one where the neighbourhood effect dominates over the pure size succession.

The need to consider the importance of size of towns has therefore been recognized, but the unresolved problem is how best to combine the joint effects of both size and distance in studying the diffusion of urban innovations, since the two have largely been treated apart from each other. On the one hand there is a wealth of literature on spatial diffusion which argues that, through various forms of communication networks whose structures are influenced by distance, innovations are likely to spread outwards from one or more originating centre in a spatially-orderly fashion. On the other hand there is a somewhat less voluminous literature which argues that, for entrepreneurial rather than household innovations, diffusion will filter downwards through the size hierarchy of cities because of the importance of minimizing risk and the economic importance of the size of market area. Representative of this latter approach is Boon's (1967) development of a diffusion model which traces the spread of information through a rank-size distribution of towns, but which does not consider the spatial distribution of those places. Similar to this is Berry's (1972) study of the diffusion of television stations in the United States over the period 1940-68. Berry uses a regression framework which equally ignores spatial aspects and concentrates on city size.[1] Having shown that the date of adoption is strongly linked to the size of the adopting city, his regression models have, as independent variables, a number of estimates of the size of the urban areas and in addition a number of measures of certain individual city characteristics which appear to be important causes of disturbance in the relationship between date of adoption and size.

The pattern of innovation diffusion through an urban system might best be thought of, however, as a combination of both the distance and the size effects. Given a particular urban innovation, its spread within a system of towns might be expected to follow both the size hierarchy and a distance spread simultaneously. Boutiques might first blossom in London and soon after spread to Birmingham, Manchester, Liverpool and Leeds, but at the same time there is a greater likelihood of their spreading to Andover, Bishop's Stortford and Chelmsford than to Aberdeen, Blyth and Carlisle. The logic for the hierarchical spread would be in terms of lessening the economic risks in a larger potential

[1] Of the independent variables, the only one which acts as a form of surrogate for distance effects is the population potential of each city. Population potential – being calculated as a spatially continuous variable – does incorporate an element of location into Berry's regression formulation, but only in the most indirect fashion.

market; and, for the 'neighbourhood' spread, the logic would be in terms of imitation of the innovation by proximate places which would have a higher probability of exposure to it. Two studies have attempted to follow this approach by combining both size and distance elements into models of diffusion in sets of urban places. Hudson's (1969) attempt takes as its starting point the formal geometry of Christaller's central-place pattern; he postulates a set of m levels of centres in which centres at the jth level are spaced apart at distances of $q^{i/2}$ where j equals $(i + 1)$ and i runs from 0 to $(m - 1)$. A town P_1 is then said to L-dominate a place P_2 through a chain of L information links. When L equals 1·0 (in other words when a message travels directly from P_1 to P_2 in only one step) P_1 is said to directly dominate P_2. The likelihood of a message passing from P_1 to P_2 is given by the expression for the demographic force of P_1 on P_2 and, since this involves both the distance between the two places and their respective population sizes, this expression subsumes the two aspects which we need to resolve together. It argues that, with respect to some originating centre which is a potential source of messages about an innovation, a place with a large population which is distant from the originating centre can be thought of as being equivalent to a smaller place which is closer to the origin. Thus the hierarchical and neighbourhood processes need not be thought of as separate and in such a central-place system one cannot think of a single type of jth-level place, since the length of the chain which links it to a potential source of messages has also to be taken into account. Hudson then assumes that each place which has heard the message about an innovation then tells every other place which it directly dominates. The process can be represented by a matrix which records the first-hearing times of places in which the columns are the order of the centre and the rows represent time. The result of the process outlined above is to produce column vectors which are binomially distributed and which therefore cumulatively produce a logistic curve of 'knowers' in the total population of places and also that the means and standard deviations of these column vectors[1] are functions solely of the parameters of the initial geometry of the central-place system – the spacing constant, q, and the number of levels in the hierarchy, m.

Pederson's (1970) model draws upon only slightly different sources since it is developed in terms of gravity ideas and the rank-size rule which give him a distance and hierarchical effect respectively. He assumes that once a place has adopted an innovation it will begin to send out messages about it to other places which have not adopted, at a rate which is determined by the gravity equation:

$$I_{i.j} = k(P_i P_j) \, . \, d_{i.j}^{-b}; \qquad [5.2]$$

where $I_{i.j}$ is the amount of information sent from place i to place j; and where P and d are population and distance respectively. The total amount of information

[1] These values gauge the respective strengths of the size effects and the distance effects: the wider the scatter and the closer the means of the columns, the less important will be the effect of town size.

received by a place which has not adopted is then simply the summation of all such flows from towns which have adopted since the time of their adoption. Thus, at time t_n, the total information at place P_i will be:

$$\sum_{j=1}^{i-1} [(k . P_i P_{j.m} . d_{i.j}^{-b}) . (t_n - t_m)] ; \quad [5.3]$$

where $P_{j.m}$ is the population of place j which adopted at date m. Towns will adopt once the fund of information has reached some critical level and will then subsequently emit messages themselves to places which have not adopted. Pederson's model thus combines the size and distance effects in a single form similar to that of Hudson. In Pederson's case, it can be seen that both the size and distance processes are limiting cases of the model which are approached as the exponent b tends to zero and infinity respectively. This is seen very clearly if the city-size distribution is approximated by the rank-size expression $P_i = P_1 . r_i^{-a}$, in which r_i is the rank of the ith town. If this expression is substituted in equation [5.2] we get the following:

$$I_{i.j} = kP_1^2 . r_i^{-a} . r_j^{-a} . d_{i.j}^{-b} \quad [5.4]$$

Here one can conceive of the ranks r_i and r_j as hierarchical distances between places, exactly equivalent to the physical distances between them (fig. 5.2); and it can be seen that as the b term is made infinitely large physical distance becomes all-important in determining the nature of the diffusion process, whereas, as it approaches zero, it is the hierarchical distance which will dominate in determining the rate at which places receive messages. The applications of this model which are pursued by Pederson suggest conclusions about the determinants of the speed of innovation diffusions and about the shape of size distributions of cities, but do not consider the consequences of this link between innovations and urban growth upon the nature of urban growth processes.[1]

We have here some most pertinent ideas about the spread of urban innovations which consistently suggest that diffusion in a set of towns might follow a path both down the levels of the urban hierarchy and spreading outwards in space from places which have already adopted. While fitting such embryonic models to empirical data has still a very long way to go, it does seem that there is some agreement about the expected general form which urban

1 On the latter he argues that once a town has adopted it will grow in size by a given percentage during each subsequent period of time. Starting from a central-place hierarchy of sizes he shows that the stepped size distribution will then be converted into a smoothed curve over time and that the shape of the curve is strongly influenced by the value of the distance term, b: with b = 1·0 a straight-line plot results; when $b > 1{\cdot}0$ a concave curve (typical of a 'primate' size distribution) emerges.

diffusion might follow. It would be characterized as a process whereby simultaneously an innovation would spread along three levels: down the urban hierarchy of sizes; out from areas of high population potential to lower population potential; and, at a more local level, outwards in space from one centre to smaller centres adjacent or close to it (Berry and Horton 1970, p. 87). This process neatly accords with Hägerstrand's (1967b, pp. 7-8) picture of the network of contacts of individuals which can be seen at national, regional and local levels, but the rationale for the hierarchical spread derives both from the flow of information and, depending on the characteristics of the innovation concerned, the degree of risk associated with different market areas. The

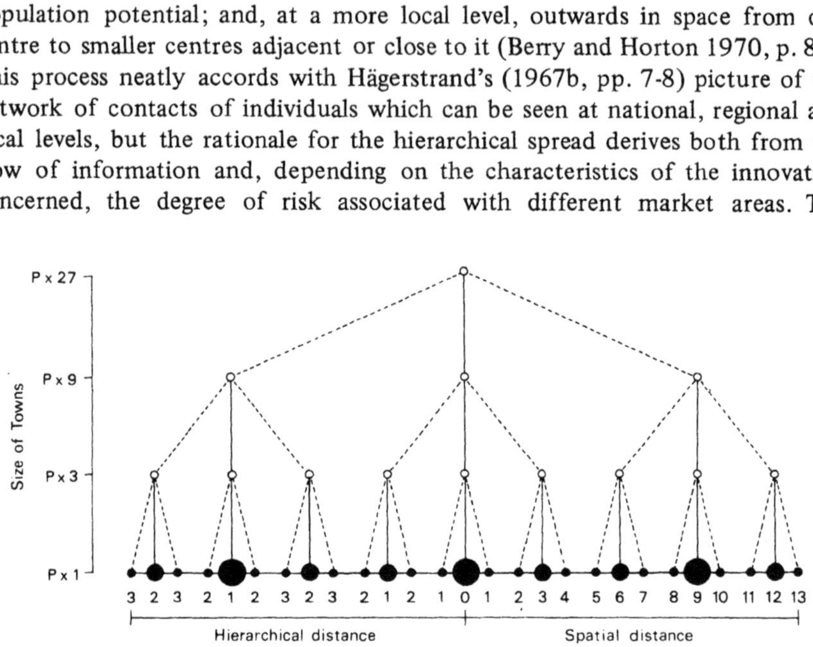

Figure 5.2 Diffusion in an urban system: hierarchical and spatial distance. *Source:* Pederson (1970)

respective importance of the hierarchical and the neighbourhood effects would therefore differ depending on the characteristics of the innovation. Those involving large economic risk or those in which technological constraints or scale economies favoured their adoption in large market areas would be determined more by hierarchical than by neighbourhood considerations.

If these ideas, albeit very roughly sketched, provide the material for an attractive model of innovation diffusion in a set of towns, to what extent do empirical data conform to the expected patterns? The difficulty in trying to answer this is, first, that tests of the fit of essentially probabilistic notions of this sort are notoriously difficult and also that, since we are trying to argue about urban growth in a nineteenth-century context, sufficiently detailed data are hard to come by. On the first point of testing the fit, this has long been a problem in all of the more well-established spatial diffusion work from Hägerstrand onwards. We have seen that Berry used a regression format to test the fit of the hierarchical spread of television stations and he found percentage explanations generally of the order of 60 per cent. This test, however, takes no account of the spatial patterning of the spread. Hudson simply tests his model in an aggregate and temporal fashion rather than at the level of individual places by looking at the curve of total hearings over time. Perhaps, in view of the problem of providing sensitive measures of fit for probabilistic models in space, one might simply look at patterns in a more purely qualitative fashion to see to what extent particular innovations appear both to show a general relationship to the size of towns and, at the same time, seem to spread out from larger centres to proximate smaller centres.[1]

On the question of data, there is a number of studies of the diffusion of technical innovations for the period of the industrial revolution. Among the more notable, with examples both from America and Britain, are studies of the diffusion of tinplate manufacture (Minchinton 1956), of Cornish pumping engines (Von Tunzelmann 1970), of iron technology (Walsh 1966), of the draper loom (Feller 1966) and of steam power (Temin 1966). But all of these studies consider the spatial dimension only incidentally. Pred (1966, pp. 106-10) provides another example in his study of the numbers of patents taken out in the United States for the three dates 1860, 1880 and 1900 in eleven cities which were to become the country's leading metropolitan centres and another five large, but declining, cities. He shows the general relationship between the numbers of patents and the size of the manufacturing population of each of the cities.[2] Perhaps the most directly relevant example, however, and one which looks explicitly at the spatial and hierarchical aspects of spread is Krim's (1967) study of the spread of urban tramways in the United States which shows very effectively the simultaneous operation of the three factors of hierarchical size,

1 If one were to test this process in a more formal manner, a regression format would appear to be most appropriate. This would involve a series of regressions for successive time periods of the form $X = a + bY + cZ + \epsilon$; where X is date of adoption, Y is population size and Z is distance from the nearest town which had already adopted by the given date. This would, however, be far from satisfactory. Once the spatial element is added, the pattern of one period becomes partly dependent on that of the previous period and any deviation – even though in itself explicable in the probabilistic framework – will tend to be amplified. The empirical data discussed later in this chapter show the way in which innovations often spread densely throughout a particular region following adoption by the major regional centre. While the whole process can be described in terms of hierarchical and spatial diffusion, a formal test of 'goodness of fit' of this probabilistic expectation would require the incorporation of so much real-world data (for example, which of the various regional centres happened to adopt early in the sequence) that the test would rapidly become circular – a problem not uncommon in attempts to test probabilistic notions in situations involving spatial 'autocorrelation' (see Pred's comments, in his postcript in Hagerstrand 1967a, where he discusses this problem in relation to Morrill's work on the evolution of settlement patterns in Sweden).

2 The fact that patents – which measure *inventiveness* rather than innovation – are taken out predominantly in urban rather than non-urban areas is discussed in, for example, Feller (1971).

population potential and distance decay which have been outlined as providing the controls of urban diffusion.

Given the very sparse number of studies which look at both spatial and hierarchical spread in sets of towns, it would be useful, before developing a model of the connection between the spread of innovations and the growth patterns of cities, to look at some examples which might illustrate the operation or lack of operation of our expected spread mechanism within nineteenth-century England and Wales. Since our argument is that technical innovations act in a cumulative fashion to boost the population growth of towns, it might appear that the best examples to study would be those which were explicitly connected with the major technical advances and industrial development of the industrial revolution. These might include the spread of the Boulton and Watt steam engine (see, for example, Lord 1923) of the late eighteenth and early nineteenth century; the spread of the major textile innovations or of those in iron and steel such as the open hearth or Bessemer processes; or the spread of the railway. Undoubtedly such innovations include the most important of what Rostow would call the 'leading sectors' behind the growth of the British economy, but they all tend to have a number of deficiencies as indicators of the way in which innovations in general might have diffused through the set of British towns. The technological innovations, for example, are specific to those particular places which practised, or which came to practise, the specific activity for which the innovations were relevant, so that the population of places 'at risk' would be specific to the particular innovation being considered. Railways suffer from a different drawback as a general indicator. The spread of railways undoubtedly had profound effects upon the growth of those towns which were part of the network,[1] as well as having more direct effects upon those places, such as Swindon or Darlington, which became major construction and servicing points for the railways themselves. But as a general indicator of diffusion amongst a set of towns the fact that railways are a two-dimensional rather than a one-dimensional phenomenon makes it difficult to characterize their spread amongst a set of towns which are one-dimensional points. The fact that a railway is built from London to Manchester, for example, means that points along the line will also 'adopt' the innovation if stations are built there, but the fact of their adoption is more contingent upon the characteristics of the terminal towns than upon their own.

Instead of looking at the major technological innovations we shall therefore

[1] For the midwestern United States in the period 1870-1900, Higgs (1969) uses multiple regression to test ideas about the growth of cities. One of his variables is the number of railway connections which a town has and he finds an 'ambiguous' relationship. He comments (pp. 374-5): 'Undoubtedly cities without rail connections stood little chance of growing substantially. But once a few railway lines had been secured, it by no means followed that procuring additional lines would guarantee further growth.' Railway connections, it might be argued, are necessary, but not sufficient elements in fostering urban growth.

look briefly at some selected examples of organizational entrepreneurial innovations. Specifically we shall look at three innovations which may, at first sight, seem to form a rather motley collection: gas works and urban street lighting; building societies and, in particular, Starr-Bowkett building societies; and telephone exchanges. While a rather odd assortment of 'innovations', all three had evident effects upon the internal functioning and, in the case of the telephone, eventually upon the external relations of nineteenth-century towns. None is explicitly connected with a particular type of town and, while none could be considered a major manufacturing stimulus in the sense of Pred's model of cumulative and circular causation, each can be considered a significant entrepreneurial innovation which would fall within the scope of his secondary multiplier effects. Certainly in the opinion of contemporaries, gas works and the telephone were considered of some considerable importance to the way in which towns functioned and, in a minor way, each would have effects upon a town's population both in terms of the direct employment which they offered and, perhaps more importantly, in terms of their spin-off effects on other types of activity – both in terms of related industrial activities and of the general attraction to migrants of towns which, having the innovations, were considered more 'progressive' or more efficient. The building societies, as will be discussed, were never a really major force in terms of the overall building activities of nineteenth-century towns. The justification for looking at the pattern of their diffusion is that, like the first two, building societies are *general* indicators of entrepreneurial innovations, but, unlike the first two, they are an innovation with very different economic implications in terms of the element of economic risk involved in their adoption.

BUILDING SOCIETIES

Building societies are not the most obvious innovation with which to examine our ideas about spatial and hierarchical diffusion. The housing market of the nineteenth century was relatively little affected in quantitative terms by houses built through such societies although Cleary (1965, pp. 286-9) estimates that as many as one in every seven or eight houses built in the 1860s were built under the auspices of or with money loaned by building societies.[1] Nevertheless, they

[1] Studies of individual towns give little quantitative indication of the relative contribution of building societies to the housing market. They do, however, suggest that they had a more than trivial effect. Church (1966, p. 263) notes that one of the hosiery manufacturers of Nottingham borrowed money from building societies to build housing for his employees towards the end of the nineteenth century. A number of the contributors to S. D. Chapman's *The history of working-class housing* (1971) also suggest the impact of the societies in specific towns: in Leeds the earliest back-to-back housing was built through terminating building societies (p. 102); in Nottingham they appear not to have played a part in early developments although 'money clubs' did (pp. 140-3); in Birmingham their effects can be shown to be of some importance (pp. 235-9); and in Ebbw Vale they had the effect of improving the standard of housing in the second half of the nineteenth century (p. 297).

did provide one source of capital for house building and their adoption may be taken as an index both of this and of the more general economic adventurousness of a town. They are also clearly entrepreneurial rather than household innovations in that the decision to start a society had in the nature of things to be undertaken communally and in the light of expectations about a sufficient number of individuals being interested to provide a relatively large and continuing flow of money into the funds from which loans were made. The way in which the knowledge of building society objectives and methods was spread also ties in with our ideas about the importance of communication networks, especially in the early years of the growth of the societies. Cleary (1965, pp. 9-10), for example, suggests that they developed out of the friendly societies which made provision for hardship and illness in a time of some general financial insecurity. The earliest of the building societies usually met in local inns and drew for their membership upon local patronage.[1] In such circumstances the spread of knowledge of the societies might be seen as being not only through the solicitors, accountants and other professional classes, who, in the early years, occasionally and, later, usually acted as the organizers, secretaries or initiators of the societies, but also through the face-to-face contacts of the members who participated in what were often relatively informal organizations. To this extent one has a nice interplay between the size-of-market aspects determining the potential number of members on the one hand and the private communication fields which provide the rationale of spatial diffusion models on the other.

Easing the provision of housing for the less affluent was the underlying, if not always the real, motive in the foundation of building societies. The elements of gambling and investment which were widespread in the later years of the century were not always absent in the early years, but were certainly far less frequent. And even though few of the societies drew upon the really poor for their membership, they did provide a means of saving and thrift and, for some, of access to housing amongst the relatively better-paid of the artisan classes and of what would now be called the lower ranks of the middle class. The ostensible aims of the societies always included the provision of housing or of capital for housing, but the means by which these objectives were furthered were very varied and underwent considerable changes as the century progressed. The early societies usually worked on some variant of a plan involving the payment of subscriptions (of the order of 10s per month – no mean sum in a population a

[1] The continuation of the connection with local inns is seen, even at late dates, by the names of many of the societies which were derived from the place where they met. Societies founded in the north in particular suggest the frequency with which inns were the meeting places especially of the terminating societies: in Bury there were the following building societies – the British Oak Inn, the Stanley Arms Inn; in Huddersfield, the Fleece Inn, the Queen's Head Inn, the Ramsden Arms, the Sir John Falstaff, the Temperance Hotel and the White Hart Inn; in Manchester, the Salford Arms, the Mitre, the Golden Oak amongst others; in Oldham, the Hart and Feathers, the Black Swan Inn.

large proportion of which earned less than 20s per week)[1] until the fund was large enough to have provided each shareholder a house, at which point the society would be broken up. They were thus all *terminating* societies which, even if successful, had relatively limited life-spans. Some of the societies themselves arranged the building of the houses. There were obviously many difficulties involved in such a scheme. First, the process could be a very long one since, as Cleary notes, with twenty members paying 10s a month it would be almost ten months before a first advance of £100 could be paid and almost as long before the second £100 advance would be possible even though later, with the interest on swelling funds, advances would come more rapidly. Second, the fact that fixed subscriptions had to be paid until the termination of the society made them very inflexible as regards membership. After the early years, the heavy backlog of subscriptions made it almost prohibitive for new members to join existing societies so that additional members could only readily be incorporated by the founding of successive societies so that the funds of a second society could be augmented from the reserve funds of the first (Cleary 1965, p. 56). Thus one finds long sequences of offspring societies formed in cascade from their parent: the first of a society called the Birmingham Building Society was formed in 1842, a second and third in 1845, a fourth in 1846 and a fifth in 1848; the first of a Manchester society, the Manchester and Salford Building Society, was formed in 1843, a second in 1845, a third in 1849, a fourth in 1851 and a fifth in 1853 (Brooks' *Directory* 1855). A third problem was connected with the provision of housing by societies where they built houses themselves. Problems inevitably arose in providing identical houses for members who had subscribed, or were to subscribe, equal amounts; the member getting a house early obviously had a great advantage over a member who got one much later. The first of these problems could, of course, be met by the society borrowing money, which helped to accelerate the process of providing returns for the subscriptions. The third problem was eased by the societies disengaging themselves from the provision of housing *per se*, but, even with the relative security afforded by the legal recognition of building societies under the 1836 Act, the terminating societies were still faced with considerable problems in ensuring regular payments of subscriptions and in equalizing the returns to their members. It was out of such problems that the idea of permanent societies grew, in the 1840s. The permanent society aimed to enable members to take out shares at any time during its existence and to give the borrower of money a fixed period over which to make repayments so that the total sum to be paid could readily be calculated. Thus advances for fixed terms of five, ten, fifteen years

1 Estimates of weekly wages taken from Baxter's 1867 classification in *National Income* are given in Best (1971, pp. 94-9). The same source suggests that the average annual income-per-head in England and Wales was £32. Figures such as these must, of course, be treated with some circumspection.

and so forth could be offered in place of societies having to demand subscriptions for an indefinite period until it wound up. Cleary (1965, p. 48) and Price (1958, p. 129) differ in their attribution of the originator of the permanent principle, but the fact of the rapid growth in the popularity of permanent societies in the period from 1845 when the Metropolitan Equitable was formed is beyond dispute.[1]

If the introduction of the permanent principle and the growth of permanent societies was the first major change in the rapidly altering aims and principles of the building society movement, the second was the more conscious introduction of the element of gambling and pure finance which came in the 1870s and 1880s. Gambling had always played a part in the movement as can be seen from the early tontines. Cleary (1965, pp. 10-11) quotes the very early case of the St John's Street Tontine formed in Swansea in 1791. It had 100 members who subscribed for five years to build five houses which were then to be let. The income from these houses was to be shared amongst the surviving members until their number was reduced to ten. At that point the houses were to be sold and the proceeds divided between the ten remaining members, or they were to continue to be let until only five members remained and the houses would be allocated to them by casting lots. Ballot societies, which grew and flourished much later in the nineteenth century, and of which the Starr-Bowkett society was a representative type, made appeal to the same sort of gambling spirit as such early tontines and in certain instances were far removed both in spirit and in practice from direct involvement with housing in any way.

Briefly this sets the background to the early developments of the building society movement. In looking at the way in which it spread through time and over space, one is severely handicapped by the intractible nature of the data. By their very nature many of the societies were of very short duration. The average life of the terminating societies may have been of the order of ten years, but many were wound up very soon after their foundation. The records of societies were also very haphazard as the Royal Commission on Friendly and Building Societies (*Second Report* 1872) found when it tried to collect material on their growth and operation. There was no central register of societies and, even though the local clerks of the peace in England and Wales held copies of the rules of the societies once they had been certified, the information supplied by the clerks to the Royal Commission was very inadequate since it failed to distinguish between the dates of foundation of new societies and the alteration in the rules of existing societies. This is a recurring and serious problem in trying

[1] The spread of permanent societies indirectly helped open the way for the development of Freehold Land Societies which grew out of the agitation for reform of the suffrage in the 1840s and aimed to help people to buy sufficient property to qualify for parliamentary voting rights. The connection of influential figures such as Cobden and Bright with these aims in turn helped to give respectability to the principle of permanence upon which the land societies operated.

Table 5.1 *Early building societies*

(*a*) *Geographical distribution*

Region	*Date of first society*	*Societies*
West Midlands	1775	Birmingham (1775, 1781(3), 1785(3), 1791(3), 1794, pre-1795); Dudley (1779, 1786, pre-1795(2)); Rowley Regis (1792); Wednesbury (pre-1820); Netherton (1821); Kidderminster (1822)
West Yorkshire	1785	Leeds (1785, 1787, pre-1814); Horbury (1793); Bingley (1806); Mirfield (1806); Skipton (1806, 1807, 1823); Bradford (1823, 1824, 1825(6)); Slaithwaite (1825)
South-east Lancashire	1792	Droylesden (1792); Cheadle (1806); Manchester (1817, 1820(2), 1821(2), 1822, 1823, 1824)
North Lancashire	1793	Preston (1793(2), 1822); Longridge (1798, 1822); Burnley (1799, 1815, 1820)
Potteries	1807	Burslem (1807); Tunstall (1816); Leek (1824)
Scotland	1808	Glasgow (1808); Selkirk (1815); Kilmarnock (1824)
London	1809	Greenwich (1809)
West Lancashire	1818	Liverpool (1818, 1825)
Cumbria	1824	Carlisle (1824)
North East	1825	Gateshead (1825(2))
North Wales	1825	Caernarvon (1825(2)); Bangor (1826); Pentir (1827); Eryri (1827)

(*b*) *Dates of formation*

	Pre-1780	*1785*	*1790*	*1795*	*1800*	*1805*	*1810*	*1815*	*1820*	*1825*
Number of societies	2	6	2	12	2	0	8	3	8	27

Note: Numbers in brackets after dates in table (a) show cases where more than one society was formed in a single year.
Sources: Price (1958), Cleary (1965), Beresford (1971).

to assemble material on the date of foundation of societies. For example, after the introduction of permanent societies, many of the terminating societies transformed themselves into permanent ones. The Home Office list of societies collected in 1859 (*Parliamentary Papers* 1859, XXIII)[1] shows a number of

[1] Fewer than 200 societies are recorded since there was no compulsion upon societies to make returns.

societies which changed their status in this way in the period of the early 1850s. The Great Grimsby and North Lincs Permanent Building and Investment Society had originally been founded as a temporary society in 1847 and became permanent in 1851; the Deal Walmer and East Kent Society was founded as a terminating society in 1846 and became permanent in 1852; the Industrial and Mutual Society of London was formed as terminating in 1852 and became permanent in 1857. In using the date of foundation given in various sources, the foundation date given may sometimes be either the date of the original foundation or the date of a change in the composition, or perhaps simply the name, of a society.

We can, nevertheless, make some cursory study of the spread of societies by using data on the dates of foundation gathered from a number of sources. Those used here are primarily the various directories issued from the 1850s onwards by Brooks, Rouse, Kent and Braund and the *Building Societies' Gazette* and the returns issued in Parliamentary Papers by the Registrar of Friendly Societies with whom a majority of building societies was registered. By amalgamating the various listings and dates of foundation provided by these sources one can derive, for the middle period of the 1850s – a time at which the movement had spread widely throughout the country – what would appear to be a reasonable estimate of the distribution of the societies at a point in time.[1]

For the very early period after the foundation of the first society – Ketley's Building Society founded in Birmingham in 1775 – the picture is particularly hazy since, even though Price (1958, pp. 59-67) produces a list of sixty-three societies known to have been formed up to 1825 to which Cleary (1965, p. 283) adds a further six, there must have been many more than this founded in total, about which nothing can be said.[2] It can be seen from table 5.1 however, that the spread of those very early societies about which there are data suggests the very strong regionalization of the diffusion. Starting in the West Midlands, the growth of societies was restricted very largely to the industrial areas of the Black Country, the Potteries, the North West and the West Riding. Indeed, the first known society in London dates from 1809, over thirty years later than Ketley's.

[1] It would have been possible to use the dates of foundation to look at the spread of societies over time, but the data are not sufficiently reliable to allow one to place any great confidence in the patterns so produced.

[2] Details of an additional society formed before 1825 are given by Beresford (1971, p. 102). This was the Crackenthorpe Garden Building Club which built back-to-back houses in Leeds and whose articles of agreement were drawn up on 3 November 1787. That there were many other early societies for which details are unknown is suggested by references to societies from Lancashire and Newcastle being prominent in the agitation which eventually led to the 1836 Act (Price 1958, pp. 80-3). These two areas do not feature as prominently as might be expected in the list of those societies for which details are known. Similarly, in the framing of the 1836 Act, constant reference is made to societies being formed 'on the Liverpool plan' which suggests a greater early development of societies in that town than is suggested by the data of table 5.1.

While the data are incomplete one can make some tentative observations on the early pattern of spread which is revealed. First there is a distinct regional bunching which could be interpreted as a reflection of the operation of some form of 'neighbourhood' effect. This can be seen in the bunching of societies in the Preston-Longridge area in the north of Lancashire, but is perhaps most clearly seen in North Wales in the period after 1825. In October 1825 the Caernarvon Building Society was formed and in the two subsequent years four further societies sprang up all within the Caernarvon area: in July 1826, the Bangor Benefit Building Society; in February 1827, the Pentir Building Society; in October 1827, the Caernarvon Union Building Society; and, also in October 1827, the Eryri Building Society (Cleary 1965, p. 283). In the light of the very localized nature of the membership and the generally informal manner in which such societies operated, it would not seem unreasonable to suggest that word-of-mouth transmission of knowledge of their operation and of their objectives would lead to a strong distance-decay element in their spatial spread which would help to account for the pattern of very strong regional clustering amongst these very early societies. Second, it is of no little interest that, where societies 'penetrated' a region for the first time, it was usually to the major, or one of the largest, towns within the region. Not only did they start in Birmingham, but the first society in Yorkshire was in Leeds; in the Potteries the first was in Burslem; and in Lancashire, Manchester, Liverpool, Burnley and Preston each fell within the ranks of the larger cities of their respective regions and all had early societies. While not trying to read too great a significance into the data, it would appear that the twin ideas of spatial contiguity and of inter-regional space jumping to the larger regional centres would adequately explain much of the diffusion suggested by these early societies.

When we turn to later periods after the 1836 Act, the number of societies has grown considerably although the data problems are still severe. Estimating the total growth in the number of societies is made difficult by the problems of registration discussed above, and the various estimates of total numbers vary widely. Contemporary commentators give estimates which vary up to as many as 7,000 in 1867 (Scratchley 1867). Price (1958, p. 237) suggests that by the centenary of the movement, in 1875, between 6,500 and 7,000 societies had been founded. A more cautious and justified estimate, based on the figures of the 1871 Royal Commission, is that by Cleary (1965, pp. 44 and 286-9) who suggests that a total of 3,500 societies were founded between the passing of the 1836 Act and 1869 and that at any one time during this period the maximum number in existence was unlikely to have exceeded 1,500 by any substantial margin. These figures certainly accord with the contemporary directory data which, while far from complete, suggest that figures of this order of magnitude seem reasonable. The data in table 5.2(a) are based on compilations from various sources and show the dates of foundation of societies which were in existence in 1853. Given the uncertainty surrounding the material, it is better to consider

Table 5.2 *Terminating and permanent building societies in England and Wales, 1853*

A. DATES OF ESTABLISHMENT

	No. of provincial societies		*No. of metropolitan societies*		
Date	*Terminating*	*Permanent*	*Terminating*	*Permanent*	*Total*
pre-1843	18		2		20
1843	4		18		22
1844	14		19		33
1845	20	1	65	2	88
1846	46	1	56	0	103
1847	36	5	42	2	85
1848	29	4	15	5	53
1849	25	9	10	2	46
1850	38	12	10	6	66
1851	21	16	14	7	58
1852	27	12	13	3	55
1853	28	14	21	4	67
pre-1850	59	2	1	0	62
Date unknown	96	5	81	4	186
Total	478	81	367	35	944

B. GEOGRAPHICAL DISTRIBUTION

	Number of societies	
Region	*Terminating*	*Permanent*
North East	30	2
Cumbria	8	0
North West	206	13
West Yorkshire	56	8
East	16	5
North Midlands	8	2
West Midlands	48	11
South Midlands	9	0
South East	39(406)	16(51)
South	14	8
South West	29	12
Wales and Border	5	3
South Wales	10	1
Total	478	81

Notes

1. For the composition of the regions see notes to table 4.2.
2. Under the South East, the figures in brackets include the totals for London and the South East.

Sources: Rouse (1853); Brooks (1854 and 1855); Kent and Braund (1884); *Parliamentary Papers* 1859, XXIII.

simply the snap-shot picture of the total numbers of societies at a given date rather than drawing inferences about the details of the temporal and spatial spread of societies from the dates of their foundation.[1] Table 5.2(b) suggests that, by the early 1850s London's late start had certainly not prevented an eventual blossoming of societies within the Metropolis and, indeed, the 1840s saw a very marked boom in the foundation of London societies. The terminal date which has been selected, 1853, postdates the period at which permanent societies had begun to develop and their expansion after 1845, especially in the South East, is noticeable. The great majority of permanent societies was found in southern England and, indeed, while terminating societies continued well into the last quarter of the nineteenth century in the north, and especially in Lancashire, the growth of permanent societies continued to become especially marked in the southern and eastern counties.[2]

The expansion of societies over time shows a discontinuous expansion in numbers. We have already seen the rather uneven expansion of the very early decades when the initial growth accelerated only very slowly, almost stopped during the Napoleonic Wars, and then accelerated markedly in the 1820s. Likewise in the later period there is some evidence of variations in the rate of expansion – with the 1840s, 1860s and 1880s being decades of more-than-usual expansion. When we look at the distribution in 1853 (fig. 5.3), it can be seen that societies had spread widely throughout the country. The particular interest in the diffusion however is the relationship with the rank size of towns. Table 5.3 plots the spread of societies against the set of towns ranked in order of their populations in 1851. The percentage of places with societies falls regularly as one moves to successively lower ranks of towns and the average number of societies in towns with societies also falls with decreasing town size. Of the top thirty towns only two places were without either a terminating or permanent society. One was Bath and this omission is in itself somewhat misleading since no fewer than four of the societies based in Bristol included both 'Bristol' and 'Bath' in their titles. The other place was Merthyr Tydfil – an anomaly which may arise as much because of the generous definition of the population of the 'town' as because of the inappropriateness of the process of hierarchical diffusion.[3] It would appear that all of the high-ranking towns in the middle of

[1] The coverage and accuracy of the data appear understandably to be best for the metropolitan area and worst for Wales and the North East. It is difficult to believe, for example, that some of the towns in which early building societies developed – such as in the Caernarvon area or Dudley in Worcestershire – did not have further societies founded in them in the period up to 1853.

[2] The geographical distribution of permanent and terminating societies for a later date, 1873, is given by Cleary (1965, p. 46) and shows the continuing preponderance of permanent societies in the metropolitan area.

[3] On the difficulty of defining Merthyr, see Davies (1926) who uses census figures and interpolates smooth waves of population growth on the basis of patterns of migration and economic development in the area. As was noted above (p. 48), the amorphous valley towns of South Wales present quite intractible problems of definition.

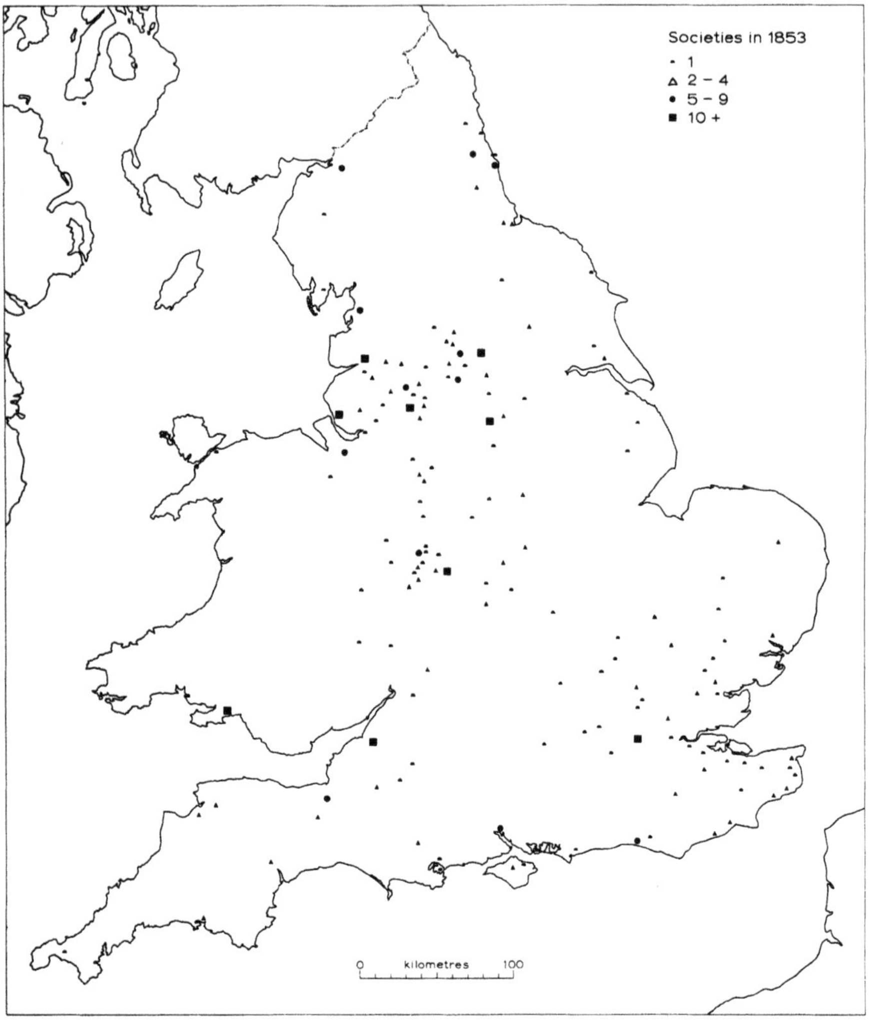

Figure 5.3 Terminating and permanent building societies in England and Wales, 1853

the century were very likely to have adopted building societies even though the total number of places with societies included also a large number of small places with only one or a few societies in addition to the large towns some of which, like Manchester with no fewer than ninety, had very large numbers. It is also interesting to note that in regions in which there is a low density of societies almost invariably one finds that the early societies in the area were in the major towns which would thus appear to lead the way in the pattern of spatial diffusion: this is true, for example, of East Anglia where Norwich had societies; of the Nottingham-Derby region where the only known societies were in the two county towns; of Leicestershire where again the county town had societies; of the Hampshire basin where Southampton and Portsmouth had societies; of the North East where Middlesbrough, Newcastle and Sunderland were represented; and in South Wales where Swansea, no doubt benefiting from its long

Table 5.3 *Hierarchical diffusion of terminating and permanent building societies, 1853*

Towns ranked by 1851 population	*Total towns in each group*	*Towns with building society in 1853* No.	%	*Total no. of societies*	*Average no. of societies per adopter*
1	10	10	100·0	402	(402·0)
2–10				223	22·3
11–30	20	18	90·0	71	3·9
31–60	30	19	63·3	57	3·0
61–100	40	19	47·5	41	2·2
101–150	50	21	42·0	41	2·0
151–210	60	19	31·6	26	1·4
211–280	70	12	17·1	17	1·4
281–360	80	16	20·0	23	1·5
361–450	90	5	5·5	7	1·4
451–521	71	5	7·0	5	1·0
Places < 2,500		27		31	1·1

connection with early forms of investment societies, had as many as eleven building societies. While such evidence is by no means as convincing as that which might be produced by a dynamic analysis which the poor data forbid, it is nevertheless suggestive of a process whereby the largest regional centres, those high in the rank order of towns, first adopt the innovation in question.

There is, too, some rather elusive evidence of the proximity effect, the second of our diffusion principles. It is this second process which provides the long tail of smaller towns which had societies. There is, for example, a particularly dense spread of societies in the South East spreading out of London. Almost all of the places in Kent[1] with populations of over 2,500 are represented by societies and the same is true of a large proportion of those towns in Essex lying close to London. Similarly, of the very small places which had societies – places so small that at no time during the period 1801-1911 did their populations rise above

[1] It was, of course, in the coastal towns of Kent that rapid urban growth rates were found during most of the decades of the nineteenth century.

2,500 to qualify them as 'towns' by our definition – an overwhelming proportion was found within the wider ambit of the Metropolis: such places included Amersham, Beaconsfield and Cheddington in Buckingham, Battle in Kent, Dunmow and Manningtree in Essex and Eaton Bray in Bedfordshire. Drawing inferences about the importance of spread effects from such cross-sectional evidence is obviously fraught with peril, but the inferences are certainly supported, frail as the ground may be. And a further independent strand of support is provided by the tendency – evident in the distribution of 1853, but much more apparent in the later spread of the 1860s – for the permanent societies, as distinct from the terminating ones, to be found much more densely developed in the areas both in and around the Metropolis than in the northern parts of England and Wales.

STARR-BOWKETT BUILDING SOCIETIES

The notion that the two principles of hierarchical diffusion and more localized spread go some considerable way in describing the diffusion of building societies can be looked at in more detail in the history of the Starr-Bowkett building societies. The history both of the originators and of the rise and fall of the Starr-Bowkett societies has the curious air of frenetic tragi-comedy which pervades so many Victorian enterprises. They boomed over a period of two decades to a growing accompaniment of dubious financial machinations and eventually were outlawed leaving a legacy which was neither undeniably bad nor demonstrably good. Bowkett certainly appears to have had the admirable intention of helping the lower-paid groups to have access to housing in his initial elaboration of his scheme and the later, and somewhat equivocal, involvement of R. B. Starr seems to have caused him some dismay. Bowkett's scheme was first advanced in 1843. It involved the formation of a society in which members paid subscriptions of 9½d per week and at the end of a period were to draw lots for an interest-free advance to be secured by a mortgage on house property, the winner of the draw subsequently paying back both his subscription and a repayment on his advance so as to clear the loan over a period of about ten years (Bowkett 1850). The continuing flow of subscriptions together with repayments would enable subsequent ballots for further advances to be held with increasing frequency until all members would have had an advance and any surplus accruing from fines and defaulting would then be repaid to the members. A certain number of Bowkett's societies were formed in the 1840s (Tompkins 1845, p. 301), but his 'Ninepenny-halfpenny' societies had only a limited success partly owing no doubt to the familiar problems of the terminating societies: the long period involved before the successful termination of the societies and the consequent temptation for those unsuccessful in the early draws to withdraw their subscriptions; and the unequal advantage accruing to those who made the earlier successful draws. Their limited success must also have owed something to the hostile reception generally given to the principle of his societies. The great burgeoning of Bowkett's principle came very much later and in the hands not of Bowkett but of Richard B. Starr who made alterations to the rules which he then copyrighted (Starr 1878). Starr's own return from the enterprise came, with increasing profit, from the fees which he charged to the local secretary and the solicitor and surveyor of the societies. The particular attraction to potential members in Starr's scheme appears to have been the opportunity which it offered for purely financial gain in a gambling context. The goal of a loan on the security of property was a valuable asset which could be sold and, as Starr altered his rules in 1877 so as to force members who had been successful in the balloting but who wished to dispose of their appropriation, to sell it to the society itself, the society could then, if it so wished, become a simple lottery in which money for housing never changed hands (Cleary 1965, p. 107). Increasingly, gamblers rather than prospective home owners appear to have played a larger part in the balloting and the increasing publicity afforded by innumerable court actions against Starr and his obvious financial success in the scheme attracted imitators in the form of a rash of other balloting societies which grew rapidly in the 1880s until, with the Building Societies Act of 1894, balloting was made illegal in future societies and the further formation of new balloting societies was abruptly checked.

In tracing the spread of the Starr-Bowkett societies, details of which have been taken from directory sources and from Parliamentary Papers, one is helped enormously by the fact that, since they were centrally controlled under the copyright of Starr, as societies were formed their sequential number was incorporated into their title. It appears to be impossible to discover the exact nature of the founding of the societies since, although copies of Starr's rule books exist, no record has been discovered of the procedure by which he founded or attracted new societies.[1] It is reasonable to assume that he himself must have canvassed for the formation of new societies as well as being approached by potentially interested groups wishing to start societies. Whether this involved his travelling from place to place for custom is not clear although it appears to have been not unlikely. So successful was his business in the 1880s indeed that he claimed to have employed fifteen people to aid him in the promotion of societies (Cleary 1965, p. 107).

If we take it that the number allotted to each society is the sequential order of foundation of that society the difficulty of attributing a precise date to its formation is obviated. This has great advantages since there is some difficulty in fixing such dates since not all societies were registered with the Chief Registrar

[1] For example, a pass book of the 433rd Starr-Bowkett Society, registered in 1881 and founded in Southampton, is in the Southampton City Record Office (ref. D/PSR/66/1). Papers of either Starr or Bowkett have not been discovered.

of Friendly Societies and, of those which were, not all were registered immediately after their foundation. Starr claimed that the first society was formed in London in 1862 (Cleary 1965, p. 105) and the subsequent progress seems to have been very rapid: the first 150 societies seem to have been founded by 1874 and there was a spate of registrations after the 1874 Act so that at least 200 had been registered by 1876 and over 700 by 1885.[1] Starr claimed that the 999th was formed in 1891. We are, however, less interested in the impossible task of trying to trace a precise time-curve of formations than of plotting what one can of the spatial spread of the societies and for this purpose the numbers of the societies have been used. Taking the first 600 societies to be formed – a number which covers the period over which the majority of new places which

Table 5.4 *Data on Starr–Bowkett building societies*

Society numbers	*Total with data*	*England and Wales*		*Scotland and Ireland*
		London	*Provincial*	
1– 50	26	26	–	–
51–100	28	27	1	–
101–150	29	22	7	–
151–200	41	21	20	–
201–250	49	21	28	–
251–300	50	19	31	–
301–350	49	11	38	–
351–400	49	15	34	–
401–450	47	9	35	3
451–500	46	10	35	1
501–550	50	17	30	3
551–600	50	13	34	3

Sources: Kent and Braund (various dates); *Parliamentary Papers* (various dates); Building Societies Gazette (1874).

formed societies had founded their first society – details have been traced for all but eighty-five of the societies. Seventy-four of those which have not been traced were in the early period up to the first 160 societies at a time when, as table 5.4 shows, the majority of societies was still being formed in London. One can be reasonably confident that, since our interest is in the date of a town's first society, these omissions are less serious than might at first sight appear. The societies are plotted in figs 5.4 to 5.9 which show the locations of existing and newly-created societies in sequence from the first 200 up to the 600th society.

[1] These figures are based on entries in the directory sources and in *Parliamentary Papers*: the directory for 1874 notes the 144th society; the date of registration of the 212th is given as 1876; and Kent and Braund's 1885 directory notes the 734th society.

Table 5.5 *Hierarchical diffusion of Starr–Bowkett building societies*

Towns ranked by 1881 population	*Total towns in each group*	*No. of places adopting up to the Nth society*					*% of places adopting up to the Nth society*				
		200th	*300th*	*400th*	*500th*	*600th*	*200th*	*300th*	*400th*	*500th*	*600th*
1– 10	10	3	7	9	10	10	30·0	70·0	90·0	100·0	100·0
11– 30	20	3	7	14	18	18	15·0	23·3	70·0	90·0	90·0
31– 60	30	6	12	18	23	25	20·0	40·0	60·0	76·7	83·3
61–100	40	3	10	18	23	24	7·5	25·0	45·0	57·5	60·0
101–150	50	2	11	16	21	28	4·0	22·0	32·0	42·0	56·0
151–210	60	–	1	5	10	12	–	1·7	8·3	16·7	20·0
211–280	70	–	3	5	11	17	–	4·3	7·1	15·7	24·3
281–360	80	2	3	3	5	9	2·5	3·8	3·8	6·2	11·3
361–450	90	1	2	3	6	10	1·1	2·2	3·3	6·6	11·1
451–550	100	1	1	3	5	6	1·0	1·0	3·0	5·0	6·0
551–650	110	–	–	2	2	3	–	–	1·8	1·8	2·7
651–749	89	–	–	–	–	3	–	–	–	–	3·4
Places less than 2,500		–	1	2	4	7					

The two processes of hierarchical filtering down the ranks of the size array of towns and of outward spatial spread to local towns are clearly apparent. Some idea of the strength of the hierarchical filtering can be got by considering the relationship with the size rank of towns in 1881. Table 5.5 shows that the pattern of diffusion follows the size hierarchy even though there is a wide scatter of early adoptions in small towns. The percentages of adoptions falls with some regularity as one progresses to groups of smaller towns. Equally, the average number of societies in adopting towns declines with smaller town sizes (table 5.6).

Table 5.6 *Average number of Starr-Bowkett societies for towns adopting by the 600th society*

Towns ranked by 1881 population	*No. of towns with society*	*Total no. of societies*	*Average no. of societies*
1	1	209	(209·0)
2– 10	9	52	5·8
11– 30	18	49	2·7
31– 60	25	45	1·8
61–100	24	35	1·5
101–150	28	37	1·3
151–210	12	15	1·3
211–280	17	20	1·2
281–360	9	10	1·1
361–450	10	10	1·0
451–550	6	6	1·0
551–650	3	4	1·3
651–749	3	3	1·0
Places < 2,500	7	7	1·0

The succession of maps suggests the spatial sequence of this diffusion. Up to the 100th society, the spread consisted of a multiplication within London and the one society found outside, the 94th, was formed in nearby Kingston-on-Thames. The spread out into the provinces began with the second hundred societies; with the 141st at Northampton, the 157th at Southampton, the 169th at Hanley in the Potteries, the 178th at Coventry, the 186th at Norwich, the 180th at Colchester, the 193rd at Great Yarmouth, and the 198th at Liverpool. All of these places were the largest or large towns within their regions. However, in addition to this spill-over down the ranks of the regional hierarchy, the density of societies within the metropolitan area as a whole shows a marked thickening. Not only did societies multiply within London, but many of the smaller proximate towns adopted the Starr-Bowkett innovation. The list includes not only the larger places such as Brighton, Maidstone, the Medway towns and

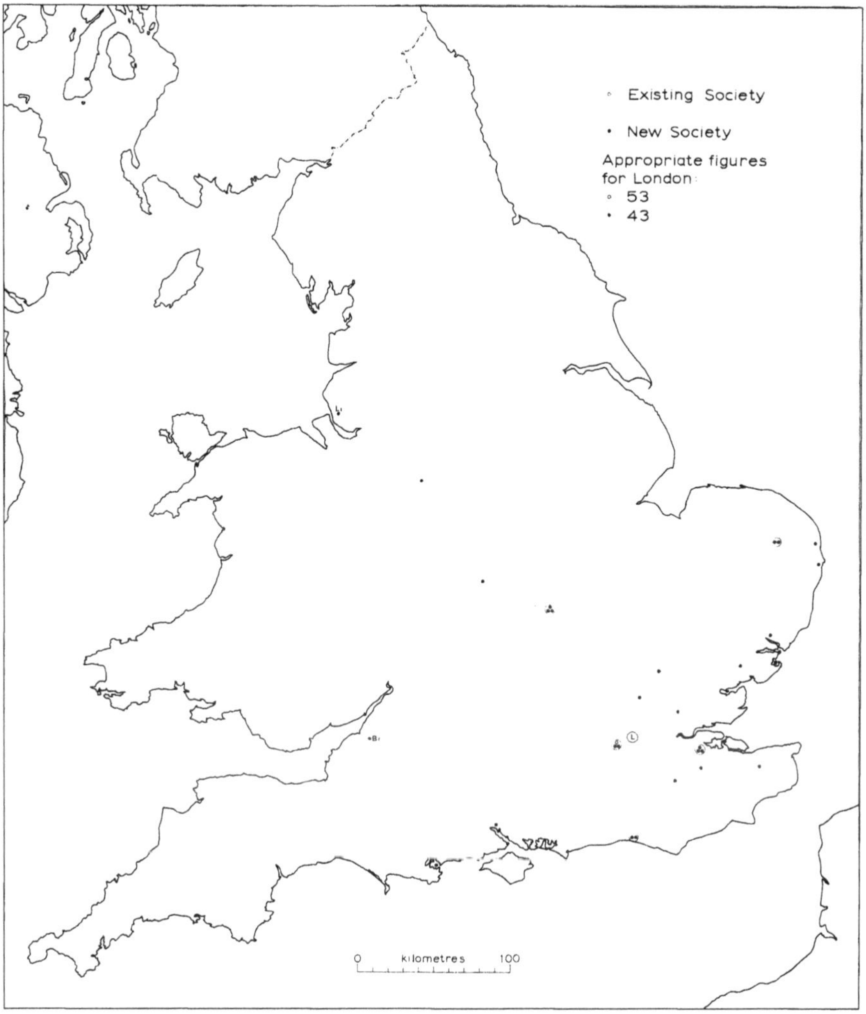

Figure 5.4 Starr-Bowkett societies in England and Wales: spatial spread of the 1st to the 200th society

Canterbury, but also a further two societies at Kingston and new ones at Brentwood, Bishop's Stortford and Cheshunt.

In the sequence of societies from the 201st to the 250th, the same two simultaneous processes can be seen. New regional centres make their appearance; Birmingham, Sheffield and Manchester. At the same time outward spread within already-colonized regions either continues, as in the south-east and eastern regions, or, as is most clearly seen in the Potteries, new clusters of local towns

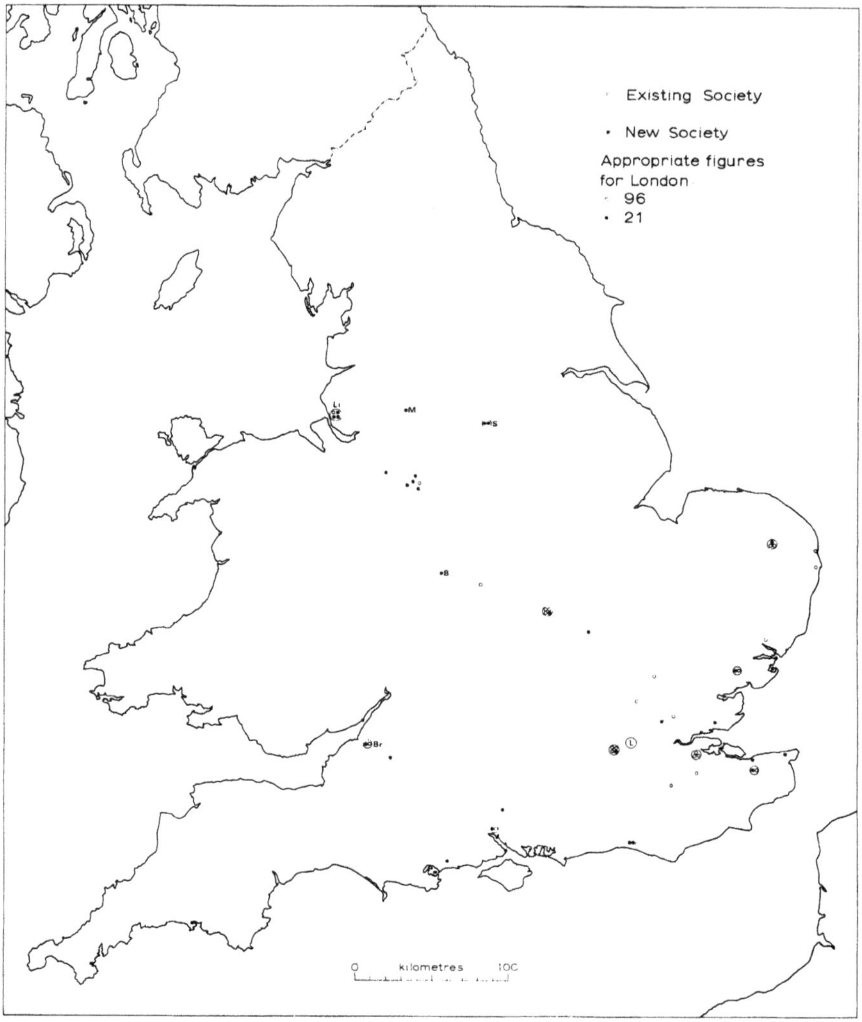

Figure 5.5 Starr-Bowkett societies in England and Wales: spatial spread of the 201st to the 250th society

adopt following the lead of a regional centre. In the period of the 251st to the 300th this process of local infilling is taken up in Lancashire and in the Black Country with new societies being formed in the smaller towns to join the established ones in the regional centres of Liverpool, Manchester and Birmingham. At the same time as this localized spread, hierarchical diffusion can be seen in the appearance of the first society in yet a further region – in the textile towns of the West Riding – with the founding of two societies in Bradford. In

Figure 5.6 Starr-Bowkett societies in England and Wales: spatial spread of the 251st to the 300th society

the next sequence, from the 301st to the 350th societies, the process of colonization within Lancashire continues and further new regions are freshly opened-up with societies being formed in the regional capitals. The North East first appears with Sunderland in the coalfield area and Stockton in the Teesside area; South Wales appears with Cardiff and Newport; and the south-west peninsula is first colonized, not through Plymouth or Exeter, but the third-largest of the Devonshire towns, Torquay, which lay in the midst of the belt of

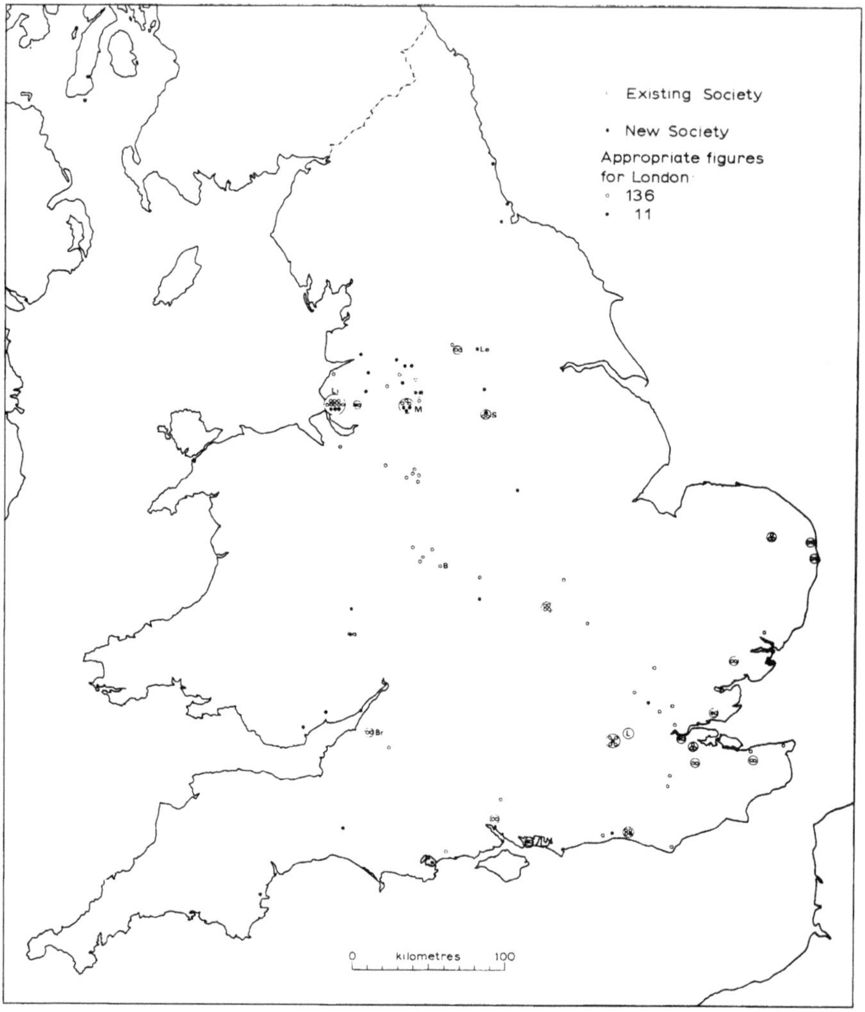

Figure 5.7 Starr-Bowkett societies in England and Wales: spatial spread of the 301st to the 350th society

seaside resorts which were the fastest growing towns of the otherwise demographically-stagnant peninsula. The later developments up to the formation of the 600th society continue to illustrate the operation of these twin processes. In particular, the development of new societies in the South West during the 'period' from the 450th to the 500th society is very striking. First, in Somerset, societies were developed in rapid succession at Chard (the 456th society), Crewkerne and Yeovil (465th) and Wells (470th); then Devon was equally

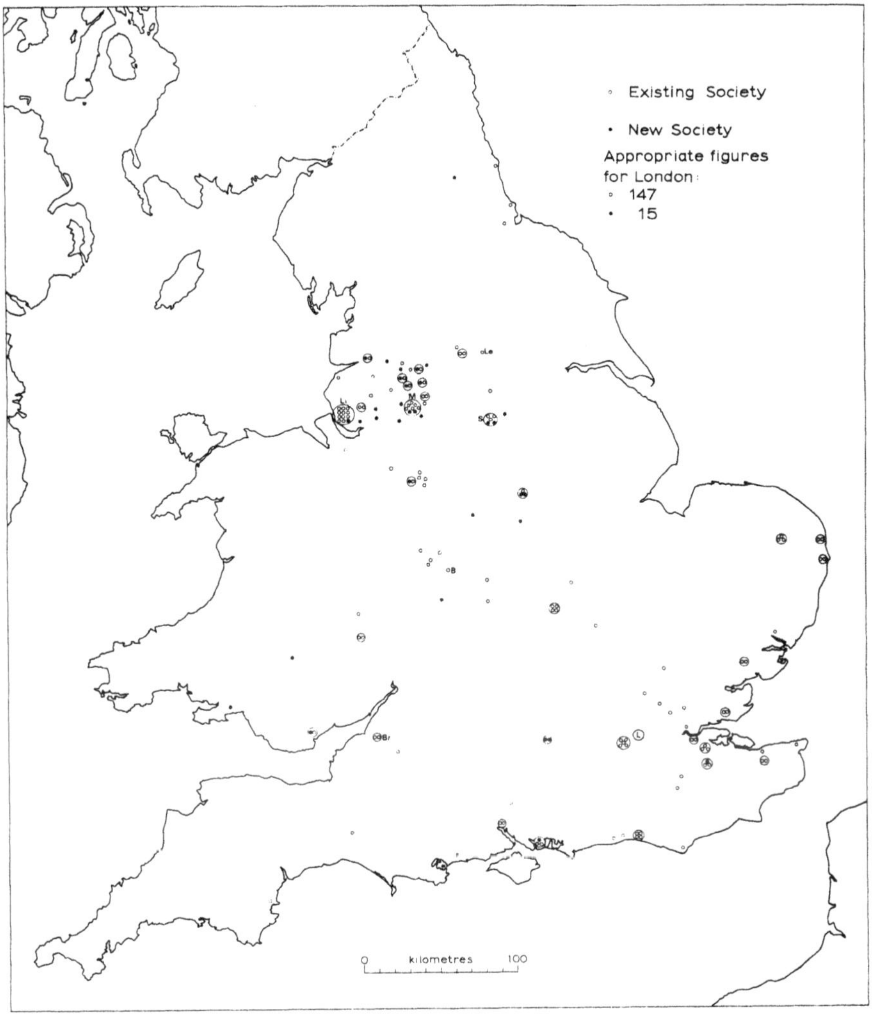

Figure 5.8 Starr-Bowkett societies in England and Wales: spatial spread of the 351st to the 400th society

rapidly colonized, with societies at Exeter (486th), Paignton (495th), Torquay 2nd (496th), Plymouth (498th) and Exmouth (499th). Developments in the North East in the 'period' between the 400th and 500th societies, likewise illustrate this same neighbourhood process. Following a lag after the foundation of societies in Sunderland (339th society), Stockton (340th), and Hartlepool (344th), others followed in fairly rapid succession: Jarrow (402nd), Sunderland 2nd (417th), Middlesbrough (421st), South Shields (436th), Middlesbrough and North

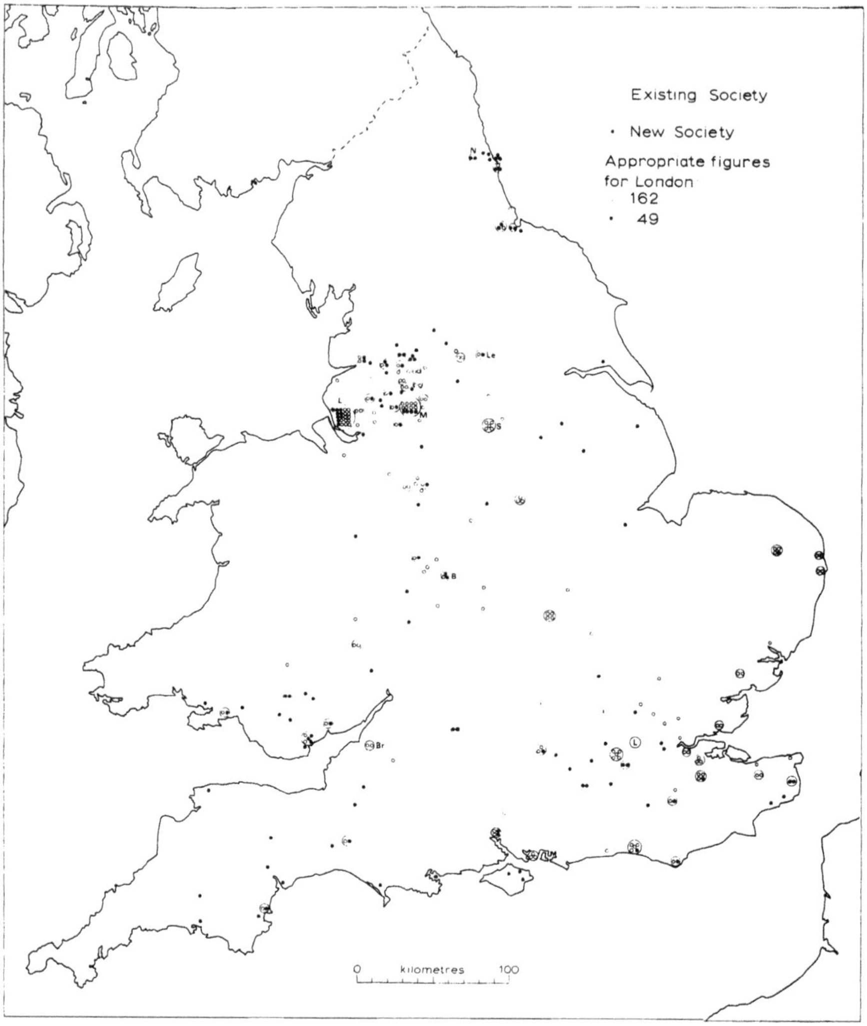

Figure 5.9 Starr-Bowkett societies in England and Wales: spatial spread of the 401st to the 600th society

Ormesby (438th), Tynemouth (452nd), South Stockton (454th), South Shields 2nd (462nd), Gateshead (477th), Stockton 2nd (481st), and Newcastle (489th). The development of Starr-Bowkett societies in the West Riding of Yorkshire was markedly lower than might be expected. This was a result of competition from the plethora of similar balloting societies – the Richmond, Model, Perfect Thrift, Mutual and others – founded in the 1880s after the success of the Starr-Bowketts and which were strongly developed in the area.

The spatial pattern of spread of the Starr-Bowkett societies is therefore a remarkably consistent one. Spread occurs first by the filtering down of societies to the regional centres or the largest towns within regions and subsequently societies spread out to other, and smaller, places within the region while at the same time consolidation occurs within regions which have already been colonized by this process. Looking at England and Wales as a whole, the two processes of hierarchical filtering and spatial spread occur simultaneously, but within any region the one precedes the other.[1] While it appears to be impossible to establish whether these societies were formed purely on the initiative of people within the various towns concerned or whether Starr or his assistants travelled from one region to another to drum up support for the scheme, the significance of these patterns remains unaffected. If the genesis of the societies was local it can be argued that the combination of hierarchical filtering and local spread would appear to be the way in which information about innovations would tend to travel in the commercial context of the nineteenth century. If people considered the Starr-Bowkett scheme attractive and feasible, the spread of formation of societies was therefore dependent upon knowledge of the idea or upon contact with its successful operation elsewhere and the probabilities of these events would seem to be explained well by our joint diffusion processes. If, on the other hand, societies were formed by Starr himself, on proselytizing missions to the provinces, we can then say that, taking him as representative of business or commercial entrepreneurs in general, such entrepreneurs must have evaluated the commercial prospects of general innovations such as this in terms consistent with our notion that innovations will diffuse simultaneously down the hierarchy and outwards from large regional centres to smaller local towns.

TELEPHONE EXCHANGES

The invention and spread of the telephone were events of considerable commercial significance to urban development. In looking at the adoption of this innovation it is clear that one is on much firmer ground in arguing that the adoption or failure to adopt could have had direct economic consequences for individual places. Of the commercial significance of the telephone there is little doubt. As Preece (1882) commented in the early 1880s, 'In 1877, it was a scientific toy; it has now grown to be a practical instrument.' It was, moreover, an instrument in which manufacturers and business men expressed great interest from the outset. Many of the early experiments in telephony were conducted on business premises. Experiments with the Edison carbon telephone in 1878, for example, used one of the longest private lines then in existence owned by the mustard firm of J. & J. Colman which linked their works in Norwich with their

[1] The same process is true of Scotland and Ireland to which societies spread rather late. Again the regional centres lead the way: in Scotland, Edinburgh formed the first societies (the 415th, 493rd and 548th) and was followed by Bo'ness (543rd); in Ireland it was Belfast (the 428th, 447th and 596th) to be followed by Ballymena (515th), Dublin (568th) and Derry (591st).

offices in Cannon Street in London (*Electrician* 1, 1878, 302).[1] The early plans to establish exchanges were usually greeted with great enthusiasm by the business communities of the towns concerned, as, for example, in the proposal to establish exchanges in Cardiff and Newport which was 'being heartily taken up by the merchants of both places' (*Electrician* 6, 1881, 149). In Glasgow – a town in which the telephone was to become more comprehensively developed than probably any other in Britain – the exchange opened by the company of D. & G. Graham established separate medical, legal, commercial and stock-brokers' exchanges and to them added a manufacturers' exchange in the early 1880s (*Electrician* 6, 1881, 98-9). In the internal organization of urban areas, the telephone was also considered of great value. Many of the various offices of town councils were connected by telephone to plant under municipal control at early dates, no doubt with the hope and the effect of improving the working of the various activities of the internal government of the urban area. Wolverhampton, for example, constructed a line of some eight miles between the water works and the pumping station and subsequently connected this line to the water engineer's office in the town hall (*Electrician* 8, 1882, 242). Particularly in the early years of its development, it was to the business and manufacturing communities that the telephone companies looked for potential customers. In his report to the third ordinary meeting of the South of England Telephone Company in 1887, the chairman made the following observations:

> The district in which we work is not a very promising one as regards business. We have few large towns, Brighton being our largest. I may say we have no manufacturing or commercial centre to compare with Bristol, Birmingham, Liverpool, Manchester, Newcastle and Glasgow. Reading, and its biscuit-baking industry and Norwich, with its mustard manufactory, are our best. Our district is composed mainly of residents and agriculturalists; the former do not require the telephone, and the latter, I much fear, cannot afford to pay for them. (*Electrician* 19, 1887, 173)

Claims and counter claims for having invented the telephone dragged through American courts well into the late 1880s (*Electrician* 22, 1888, 35) since in the 1870s there was a variety of types of equipment used in telephony some, but not all, of which used different principles for their receivers and transmitters. Whatever the date and whoever the inventor, there seems little doubt that the telephone was first introduced to Britain in a demonstration at the British Association meeting in Plymouth in 1877 (Preece 1882) and that it was rapidly promoted by the formation of London- and provincial-based companies founded in the late 1870s. The two principal companies – The Telephone Company

[1] Sources of data on the development of the telephone are primarily the weekly journal *The Electrician* and the papers collected in the archives of the General Post Office. The former is referred to in the text as *Electrician*, followed by the volume, year and pages; the latter is referred to as *GPO* followed by the reference number of the appropriate archive.

Limited (Bell's Patents) which was floated in 1878, and The Edison Telephone Company of London Ltd which was floated in the following year – were joined to form The United Telephone Company in an amalgamation in 1880. In the provinces a variety of companies was formed both as subsidiaries of the London company and as independent enterprises. The plethora of separate companies and the, at times, extremely bitter fighting between them, was no doubt encouraged by the fact, as Baldwin (1925, p. 91) comments, that Bell's receiver and Hughes' microphone were both instruments of simple design and so readily lent themselves to simple forms of construction and to experimentation. Out of the flurry of formations of fresh companies both in London and elsewhere, a much simpler pattern of control had evolved by the late nineteenth century. Through amalgamations and take-overs of smaller companies, the 1880s were dominated by the United Company in the metropolis and a series of regionally-based companies elsewhere: the South of England Company; the Lancashire and Cheshire Company; the Western Counties and South Wales Company; the Northern District Company covering the North East, Cumbria and part of Scotland; and the National Company covering Yorkshire, Nottinghamshire and much of Scotland. In the face of the increased potential competition with the expiry of patent rights, amalgamation of these main companies was begun in 1889 by the agreement of the United and the Lancashire and Cheshire companies to be subsumed into the National and this process continued with the take-over of the other companies until, by the early 1890s the National Company stood virtually in a monopoly with its only competitor being the Post Office itself. The state finally bought out the National Company, taking over its plant in its entirety in 1912 to create a state telephone service.[1]

There is some difficulty in establishing the date of adoption of the telephone in particular towns. Ideally, the spread of adoption could best be measured by the opening of exchanges in towns, but this is complicated in a number of ways. First, the opening of exchanges is nowhere recorded formally. The companies were licensees of the Post Office and had to apply to the PO for licences to operate exchanges.[2] However, the material on licences is far from complete and

[1] The only exception to total control by the Post Office was the town of Hull which was in the hands of the local authority – the sole vestige of an earlier flurry of interest in municipal telephony. For a very readable account of the early history of the telephone see Baldwin (1925). A more general account is provided by Webb (1910).

[2] This position arose out of a court decision in 1880 that the private telephone companies infringed the state monopoly in the telegraph (see below, chapter 7, p. 225). An instructive illustration of the operation of hierarchical filtering of information through the largest places in a set of towns – a process strongly underlying much of our earlier discussion of diffusion – is provided by the way in which the Post Office disseminated news of this court victory over the Edison Company. The result of the court decision was sent by telegraph in October 1880, to local postmasters who were instructed to inform the editors of the principal morning newspapers. At the same time, they were to place an advertisement in the newspapers which drew attention to the fact that 'in order to meet the convenience of the public, the Post Office is now prepared to establish in any town a system of intercommunication by means of the Telephone instrument . . .' The towns to which these

the date of obtaining the licence was very often different from the actual opening of the exchange. The opening of exchanges in the early years of the development of the telephone were not surrounded with publicity and glamour as were later openings (Baldwin 1925, p. 116) and notices of their opening only rarely appeared even in local newspapers. Secondly, the transition from private lines to full exchange services was a gradual one. Frequently an area would have one or a number of private lines linking private houses with offices, linking a residence with its lodge or its stables, or linking different branches of a single company. Out of such private lines, an exchange would often eventually be built and full exchange services then be offered to the general and business publics.[1] Thirdly, there are many cases in which meetings of local bodies established their interest in the opening of an exchange and in which the telephone companies then promised its immediate execution, yet the exchange may not have been opened for some considerable time after such interest was first expressed. For example, a meeting to discuss the opening of a Gloucester exchange was held as early as February 1880 (*Electrician* 4, 1880, 169), yet an exchange appears not to have been opened until June 1887. Likewise, even though the town council of Derby was experimenting with telephony in 1881 (*Electrician* 7, 1881, 209), an exchange was only opened in 1888.

It is for such reasons that the date of establishment of even the very first exchange in Britain is in some doubt. Different authorities suggest London, Glasgow, Manchester – 'all that documentary evidence has produced so far is that there is very little to choose between the claims of the three exchanges . . . to priority' (*National Telephone Journal* 6, 1910, quoted in Baldwin 1925, pp. 117-18). For our purposes, rather than rely on what data have been collected on the actual opening dates of exchanges in individual towns, three dates have been selected at which to illustrate a static cross-sectional approach to the diffusion. The first date is the end of 1881, at which time the early adopters had opened exchanges despite the opposition of the Post Office which itself had begun to open its own services by converting certain of its telegraphic services to telephone circuits. The second date is the end of 1886, at which time, with a

messages were sent were selected from 'a list of towns in which daily newspapers are published receiving regular supplies of news'. Those places selected, in addition to the principal London newspapers, were: Aberdeen, Belfast, Birmingham, Bradford, Bristol, Cardiff, Cork, Darlington, Dublin, Dundee, Edinburgh, Exeter, Glasgow, Huddersfield, Hull, Ipswich, Leeds, Leicester, Liverpool, Manchester, Middlesbrough, Newcastle, Newport, Norwich, Nottingham, Plymouth, Sheffield, Southport, Sunderland, Swansea and York. (*GPO*, E.13267/1889/Post 30.) Of the twenty-five largest towns in England and Wales in 1881, only eight are not included in this list.

[1] Accompanying the annual statements of many of the early regional companies are lists of exchange and private lines open at the appropriate dates. They show that a number of towns had often established long private lines well before the opening of full exchange services: for example Huyton in Lancashire, Windermere in Westmorland and Bishop Auckland in Durham for long had a number of private lines well before the opening of exchanges; and a private line connected Alnwick and Alnmouth in Northumberland long before exchanges were opened at either place.

relaxation of the adamant refusals by the Post Office to grant licences, the number of exchanges had covered large areas of the country and the system was poised to develop a full-scale network of services by the connection of regional systems through the development of trunk lines. The third date is the end of 1892 by which time an inter-regional trunk network had been established and the telephone was beginning to spread even to very small places. The data have been taken from a number of primary and secondary sources. Of the primary sources, the records of the GPO Archives are invaluable even though surprisingly not as complete as one would have expected for so recent a period. Notes and articles in *The Electrician* have been extensively relied upon to establish dates of openings, as have the reports of the various companies to their annual meetings. For the final date, Gunston (1907-8) provides a survey which is probably as accurate as could be wished since he was himself an employee of the National Telephone Company which, at the time that he wrote, enjoyed a virtual monopoly and which doubtless enabled him to draw extensively on documentary evidence which appears no longer to exist. In addition, the telephone directories are useful supplementary sources, particularly in the period from 1896 when the National Company issued an annual directory covering the whole country. Before this date, a number of directories listing the subscribers and activities of separate companies were intermittently issued and a few have adventitiously been preserved and are in the GPO Archives.[1]

The pattern of diffusion which emerges from the compilations based on these sources produces a striking picture. For the first date (fig. 5.10) the spread is restricted almost entirely to the very largest towns in England and Wales.[2] Of the exchanges shown as being open by 1881, all of the largest towns in the West Midlands, north-west Yorkshire and the North East had exchanges, as did Bristol in the South West. London by this date had as many as fifteen exchanges (Baldwin 1925, p. 75)[3] although, as applied throughout the early history of the

[1] Directories preserved in the GPO are as follows: Northern District Telephone Company, 31 January 1886 (*GPO*, 1/6); *idem.*, July 1889 (*GPO*, 1/10); Post Office Telephone List of Subscribers to the Newcastle-on-Tyne and District Telephone system, 21 March 1887 (*GPO*, 1/9); *idem.*, 31 August 1892; *idem.*, 31 December 1892; *idem.*, 30 June 1893; *idem.*, 30 April 1896 (*GPO*, 1/12); Lancashire and Cheshire Telephone Exchange Company, January 1886 (*GPO*, 1/7); National Telephone Company, Midland District, September 1886 (*GPO*, 1/8); *idem.*, Nottingham District, 1 January 1885 (*GPO*, 1/5); *idem.*, Thames Valley District, May 1894 (*GPO*, no ref.); *idem.*, (northern England and Midlands), 1891-2 (*GPO*, 1/11).

[2] Developments in Scotland have been excluded from this study even though, in proportionate terms, they were far more impressive than those in England and Wales. The exchange opened by the company of D. & G. Graham in Glasgow in March 1879 would appear to be the first ever opened in Britain (Baldwin, 1925, pp. 117-18) and the spread of further exchanges and the growth of a trunk-line system in central Scotland were far more dense than in virtually any area south of the border.

[3] By December 1881, these fifteen exchanges served a total of some 1338 subscribers. London in fact lagged behind many of the provincial cities both in terms of the quality and extent of the telephone service. Annual meetings of the United Company in the 1880s frequently dissolved into acrid attacks on the failings of the London company by comparison with its regional offshoots.

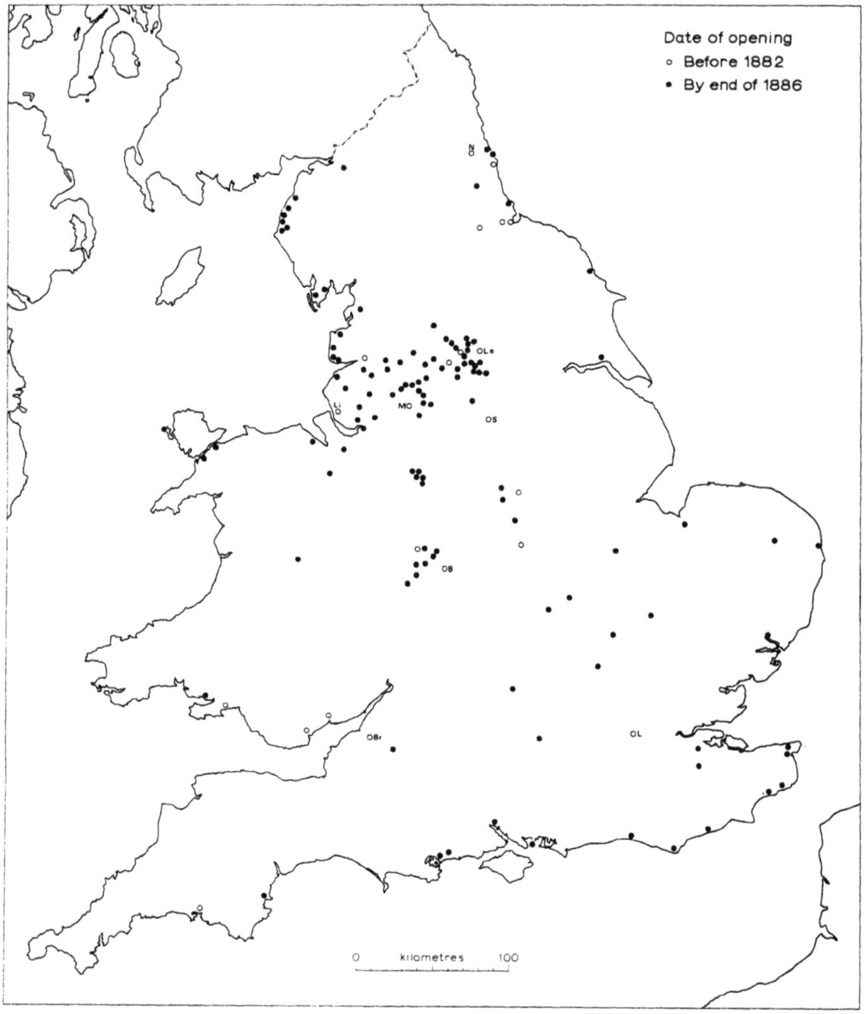

Figure 5.10 Telephone exchanges in England and Wales: spatial diffusion 1881-6

telephone in the Metropolis, exchange areas tended to be very small, and served relatively few subscribers.[1] The Newcastle and Plymouth exchanges had an interesting early development. The first Plymouth exchange had been opened in

[1] The size of area and of population served by exchanges could vary very widely; the largest being found in northern towns. In 1887, for example, within the area of the Northern District Company, Sunderland's single exchange served 215 subscribers, whereas that at North Shields served a mere 5 (*Electrician* 18, 1887, 194). Such differences add weight to the use of the date of opening of the first exchange and the exclusion of subsequent exchanges in towns where more than one were opened.

July 1881, but was then refused a licence by the Post Office and was forced to close in October of the same year and was not reopened by the company until 1885; the Post Office in the meantime having opened its own exchange in December of 1881 (Baldwin 1925, pp. 110-11). Likewise in Newcastle, a company exchange had been opened in 1881 and, being refused a licence, was forced to close in July 1882 and eventually reopened in May 1883. The Post Office started its own exchange in the town in October 1882 (Baldwin 1925, pp. 146-8). Indeed, the whole of the North East was an area of considerable Post Office activity. The first exchanges in Middlesbrough, Stockton and Darlington and, later, in Hull, were started by the Post Office and it was because the department had 'established exchanges worked both by ABCs [the telegraphic system used by the Post Office] and Telephones' (*GPO*, E.4806/1881/Post 30) that licences were refused to private companies to establish exchanges in these towns.

By 1886 the spread of exchanges had been considerable (fig. 5.10). With the establishment of the regional companies, there was a number of provincial exchanges scattered across the counties of southern England, and particularly those of the South of England's exchanges along the south-east coast, but the most striking developments had been in the industrial areas where both Lancashire and Yorkshire had a very large number of exchanges. The other industrial districts were almost as flourishing. The Potteries and the West Midlands each had exchanges in most of their larger towns, and the North East, both in the hands of the Northern District Company and, to a lesser extent, of the Post Office, had continued to expand its coverage. Only the Nottingham-Derby area and the towns around Sheffield appeared to be markedly deficient at this date.[1]

By 1892 the spread of coverage was very wide indeed, (fig. 5.11) with large numbers of new exchanges having opened in South Wales in particular as well as in the non-industrial areas of Kent and Devonshire and in the area of the Lancashire and Cheshire Company which extended to cover the towns of North Wales.

The essence of this pattern of diffusion can be seen most clearly if the dates of adoption of exchanges are again plotted against the population of the towns. This is done in table 5.7 where, for groups of towns ranked by their populations in either 1881 or, for the final date, in 1891, the numbers and proportions of places which had adopted by the three dates are shown. The fall in the percentage of adopters as one moves into lower population sizes is very regular.

[1] In Chesterfield, in Derbyshire, an exchange was eventually opened by the National Company in 1887. Just prior to its opening it was said that the town 'was probably at the moment the most important town in the country without [an exchange]' (*Electrician* 19, 1887, 469). The commercial significance of the establishment of exchanges is suggested by the importance attached at the time to the fact that Chesterfield would also be linked by trunk line to Nottingham and Sheffield, thus 'bringing not less than 10,000 firms within hail of the local exchange'.

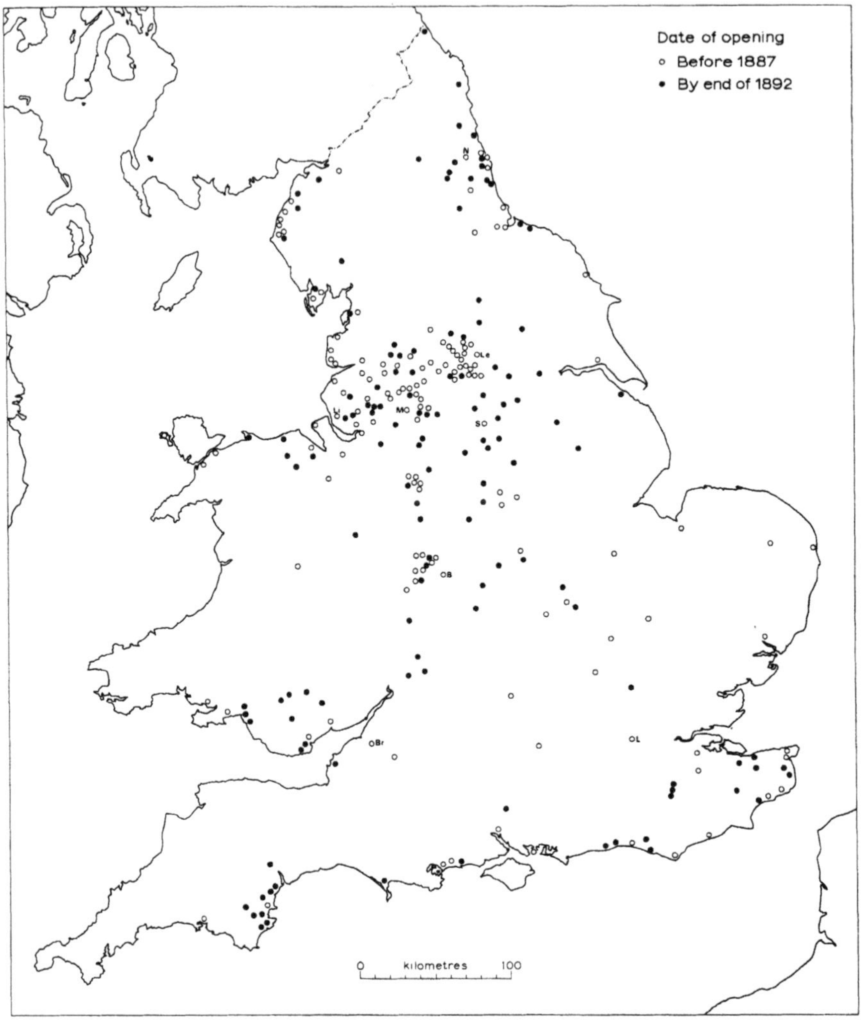

Figure 5.11 Telephone exchanges in England and Wales: spatial diffusion 1886–92

At the first date, the top ten towns had all adopted and, of the other adopters, almost all were found in the top thirty places. By 1886 all but two of the first fifty towns had adopted, but the scatter of other adopters had begun to creep very rapidly down the ranks of population size with no fewer than seven places of less than 5,000 population having had exchanges established. By 1892 this process had continued with infilling of the upper town ranks and large numbers of places in the very small population sizes having exchanges. By this final date, of the top 100 towns, only seven appear not to have had exchanges. Rhondda in South Wales was the largest of these exceptions and its anomalous position can be accounted for by the problems of defining the Rhondda as a town. Of the remaining six, all opened exchanges soon after 1892: Kingston-on-Thames and Swindon in 1893; Colchester and Crewe in 1894; Gravesend in 1895;[1] Enfield in 1899. At the other end of the size range, there was a number of places with populations of less than 2,500 (and thus not qualifying as 'towns' by our criteria) which nevertheless had exchanges by 1892. Excluding exchanges in the suburban areas around major towns (such as Moseley and Selly Oak in Birmingham, Clifton in Bristol, Kemp Town, Hove and Preston in Brighton, Headingley in Leeds) there appear to be at least twenty-eight exchanges open at that date in places smaller than this population limit.[2]

Table 5.7 *Hierarchical diffusion of telephone exchanges, 1879–92*

Rank of towns[1]	*No. of towns in each group*	*No. of places with exchange* 1881[3]	1886[3]	1892[3]	*% of places with exchange* 1881[3]	1886[3]	1892[3]
1– 10	10	10	10	10	100·0	100·0	100·0
11– 30	20	9	19	19	45·0	95·0	95·0
31– 60	30	3	22	30	10·0	73·3	100·0
61–100	40	1	21	34	2·5	52·5	85·0
101–150	50		22	34		44·0	68·0
151–210	60		13	27		21·7	45·0
211–280	70		9	30		12·9	42·8
281–360	80		6	21		7·5	26·3
361–450	90		6	20		6·7	22·2
451–550	100		5	20		5·0	20·0
551–660	110		3	12		2·7	10·9
661+	–[2]		1	12		1·1	6·9

Notes

1. For 1881 and 1886, towns are ranked by their populations in 1881; for 1892, they are ranked by populations in 1891.
2. In the final group of ranks the 1881 distribution has 89 observations, the 1891 distribution has 174 observations.
3. Dates are for the end of each year.

Sources: Baldwin (1925); Kingsbury (1915); Gunston (1907-8); *Electrician* (1878 onwards); GPO Archives.

[1] Some doubt surrounds the date of opening of the Gravesend exchange. A report of the South of England Company in 1886 claimed that a Gravesend exchange had already opened (*Electrician* 17, 1886, 270), but the report of the third ordinary meeting of the Company in the following year suggests that an exchange had not been opened by that date (*Electrician* 19, 1887, 152).

[2] These places therefore do not appear on the maps of diffusion – the suburban exchanges, as with all of the London ones, have been treated as part of the town in which they occur. See above, pp. 49–50, for definitions of certain of the largest towns.

The diffusion of the telephone exchange appears to have followed guidelines very similar to that of building societies. The same two principles of size and proximity are apparent in both. The differences lie in the relative importance of population size in providing the early and main determinant of the pattern of spread. In the case of the telephone, size was obviously of considerable importance. The potential returns from installing the various switching equipment, lines and personnel could only be realized if a sufficient number of subscribers could be found and consequently size of population offered one of the most important incentives for the establishment of an exchange. The state telegraph service had earlier provided some warning of this since, after its initial hopes of establishing telegraph offices in villages throughout the country after the state take-over in 1870, it had been forced later to demand guarantees of certain levels of usage before the Post Office undertook to open offices. Likewise with the telephone, it is apparent from the comments of company managers that it was the size, and more particularly the industrial and commercial sizes of a town which lay uppermost in their minds in determining whether or not to go ahead with the installation of an exchange. There was, however, more to the diffusion than mere size alone. From table 5.7 it is apparent that even quite small places were developed at relatively early dates. Almost invariably these unusually small early adopters were towns close to larger places which had previously adopted. Most are therefore to be found in the industrial areas of Lancashire, Yorkshire and Staffordshire. The principle of proximity operated no less clearly than did size, even though, in comparison with the spread of building societies in which the economic consequences of size were not as serious, the proximity effect is much more subdued. This localized spread effect was the consequence not so much of imitation (although the lobbying by town councils and local chambers of commerce from relatively small towns must no doubt have been prompted partly by knowledge of the successful working of the telephone in places which they knew), rather was it connected with the regional policies of the respective companies in whose hands the developments were undertaken. And the spread within a region such as the North West was very much affected by the rapid growth of a network of telephone trunk lines which connected up the emerging exchanges and which in turn affected the future establishments of exchanges. Gunston comments: 'It is not the rurality or size of a town which is most likely to make it a latecomer in the telephone network. It is rather its remoteness from and inaccessibility to the busier and more crowded areas' (Gunston 1910, p. 73). Whereas, with the development of a trunk network, very small places may have got exchanges because they lay with the line of trunk development, other places which lay off such lines often had to wait very much longer for telephones. Thus Wigton in Cumberland (with a population in 1891 of less than 4,000) had an exchange by 1892 no doubt because it lay on the line of the network of trunks connecting the towns of west Cumberland with Carlisle. On the other hand, places in the southern part of

England which lay off the main trunk lines from London to the industrial north and those northern towns remote from the main centres in the north had to wait until the very end of the century before they got exchanges even though their populations may have been quite large. The eastern part of the country suffered notably in this respect; Grantham, for example, with a population in 1901 of over 17,500, had its first exchange as late as 1900 and it was by no means an isolated example.

Earlier it was stressed that the innovations which would be studied were one-dimensional *point* activities which would illustrate the diffusion of innovations amongst a discontinuous set of towns. The development of a network of telephone trunk lines may appear to contradict this claim. However, the fact of the development of the trunk system did not make the diffusion of the telephone an inappropriate innovation for our purposes, mainly because the period which has been studied largely predated the establishment of a national system of trunks. The early telephone developments indeed were expressly denied the opportunity to develop inter-town communications since the Post Office, which licensed telephone developments, wished to protect its telegraph monopoly. The very early licences were issued for defined radii around the specific towns in question. Licences issued before 1881 included the following typical radii (*GPO*, E.4806/1881/Post 30): Birmingham and a radius of 4 miles from the central post office; Halifax and a radius of 3 miles; Wednesbury with a radius of 1·5 miles; Leeds with a radius of 5 miles. Such restrictions were universally applied wherever early licences were issued. In the early years of development, indeed, there was little real comprehension of the potential scope of long-distance telephony. No doubt some of the apparent disbelief in the long-distance potential of the telephone was not as disingenuous as might first appear since it was in the interests of the advocates of the telephone to underplay its potential competition with the state telegraph. A leading article in *The Electrician* of 1881 argued that 'the telephone is to the telegraph what the car or omnibus is to the railway . . . there is a limit to the extension and multiplication of the telephone', and concluded that the creation of a wider network through overhead wires or cables should not be attempted since it was only a local instrument (*Electrician* 7, 1881, 200). Not only, from our post-Beeching vantage point, does the comparison seem singularly ill-chosen, but, given the vacillating stance of the journal with respect to the telephone, it is also difficult to know whether such a judgement was being passed on purely technical or political grounds. Disbelief in the feasibility of long-distance telephony does, however, appear to have persisted for a considerable time. Commenting on reports of successful experiments of telephony over a distance of 200 miles in the United States, a correspondent to *The Electrician* in 1886 (16, 332-3) discussed claims that the practical working of the telephone between London and Birmingham – a distance of 113 miles – was not possible and debated whether or not this was so. Later still, in 1888 (21, 12-13) *The Electrician* commented upon the amalgamation of companies, suggesting that this was being hastened by the growth of the trunk system – 'The rise and progress of this branch of their business [the trunks] has been almost as much a surprise as the rise and progress of telephony itself . . . it is now to all intents and purposes a telegraph worked by telephony.' The conclusion was that, had this development been foreseen by the Post Office, its attitude to licensing would have been even more restrictive, given its interests in protecting the state telegraph service.

While the awareness of the potential of long-distance telephony appears, for whatever reason, to have been slow, the actual development of a network of trunk lines evolved inexorably from the start. Early hopes of inter-town links were dampened by the undoubted problems of insulation, but in the provinces long-distance trunks appear at very early dates. Leeds was connected with Bradford as early as 1880 and by the middle of the 1880s there was a number of small regional networks each of which was separate but which were eventually to be linked in to a national network. Such limited trunk developments were best developed between the major towns around Manchester, Leeds, Birmingham and Sunderland. With a new system of licences after 1884, however, trunk line developments proceeded quite rapidly and a series of networks developed which eventually were linked one to another, particularly in the areas of the Lancashire and Cheshire Company,[1] the Yorkshire area in the hands of the National Company and the North East under the Northern District Company. London was remarkably isolated by comparison. Its first inter-town link was completed only by 1884, and this only connected it with Brighton. It was not until 1890 that a line was opened connecting London with the industrial centres of the north.

Figure 5.12 shows the network which had evolved by the end of the period of our immediate interest– in 1892/3. By this time, not only was London linked to the relatively dense networks of the industrial regions, but many of the originally separate networks in other parts of England and Wales (for instance in South Wales, and the South West) had been joined in to the evolving national system. What needs to be stressed in relation to our particular interest in the diffusion of the telephone, however, is the novelty of such a national network by 1892 and the fact that, throughout the greater part of the 1880s, the development of this net had been in disconnected regional sub-networks.[2] The effects of this trunk line development were to reinforce, or provide a stronger economic rationale for, the essentially local spread of the diffusion of the

[1] Which company had, by 1887, almost one-third of its total length of wire used for inter-town connections (*Electrician* 19, 1887, 217).

[2] The differences between the developments of the telephone and the railway networks illustrate the differing importance of long-distance connection in the two. The railway network very rapidly became a national one (see, for example, the maps of its extent in 1840 and 1850 in W. Smith, 1953, p. 161), whereas the telephone was essentially a local phenomenon throughout the decade from its first introduction to the late 1880s. In making such a comparison one has to allow for the much slower overall speed of spread of the railways.

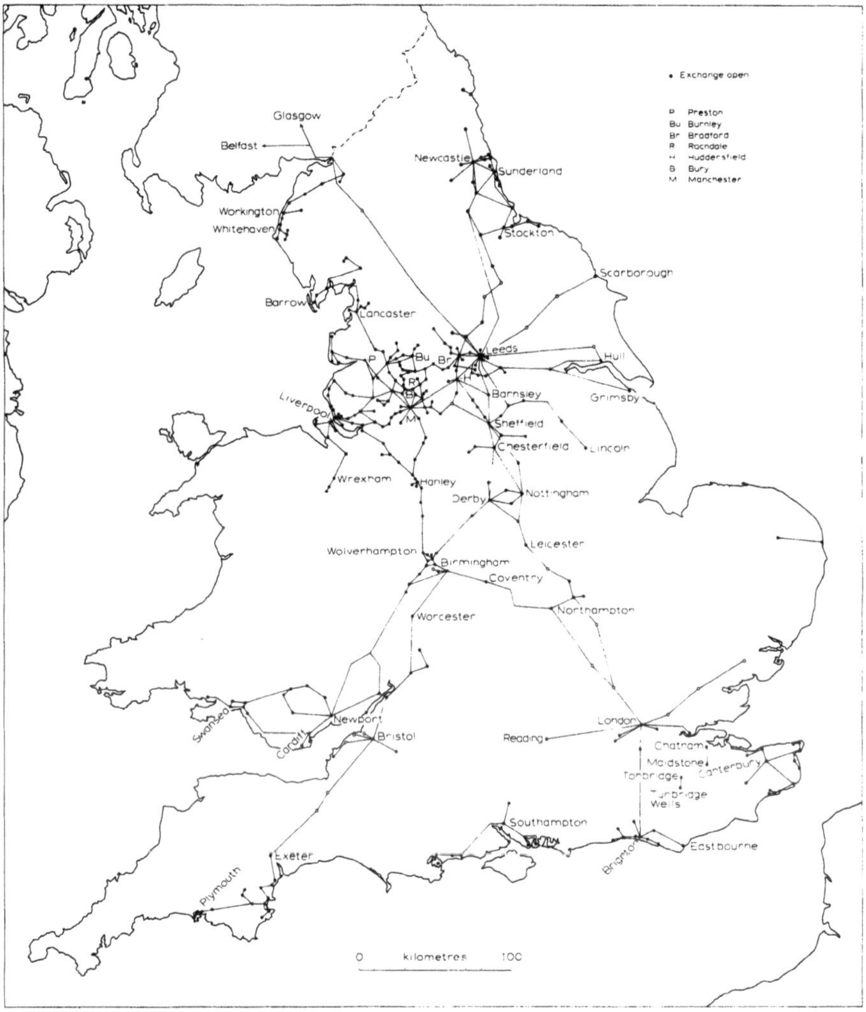

Figure 5.12 The telephone trunk-line network, 1892. *Sources:* National Telephone Company, Trunk lines 1891 (*GPO*. Ref 3/12B); National Telephone Company . . . Exchange and Trunkline system 1892 (British Museum. MAPS. 15.a.31)

telephone. Its diffusion must therefore be seen both in terms of an over-riding control being exerted by the size of town, especially in the very early years of development, as well as in terms of a local outward spread – either by imitation or through the effects of the development of a trunk network – to smaller places in the vicinity of the large early adopters.

GASWORKS AND STREET LIGHTING

The development of town gasworks must have been one of the most immediate and striking symbols of the industrial revolution. The very first company – the London Gas-light and Coke Company – was founded in 1812, yet by 1841 a contemporary was able to write that there was 'scarcely a town of any importance' which was not lit with gas (Clegg 1841). For people living in the smaller towns and especially in the non-industrial southern parts of Britain, the spread of gasworks must have been perhaps the one visible sign of the 'progress' of industrialization in the first half of the century. And the visual impact must have been the more marked because the erection of gasworks was almost everywhere associated with the lighting of streets since, in return for the right to break open the street surface to lay pipes, local authorities demanded that gas companies provided street lighting. It was doubtless this fact that gasworks provided so immediate a symbol of the industrial transformation of the age that accounts for the prominence given to them both in contemporary directories and in other general descriptions of towns in the first half of the nineteenth century.

The pattern of the diffusion of gasworks has already been discussed by Falkus (1967) who reached three conclusions which have bearing on the nature of the spread of entrepreneurial innovations: first that, by about the middle of the century, the vast majority of towns with populations greater than 2,500 possessed gas companies; second, that the construction of gasworks took place in surges which corresponded closely with the trade cycle; and third, that throughout the period there was a marked tendency for gasworks to diffuse from larger to progressively smaller towns. As a general indicator of the diffusion of entrepreneurial innovations, the gas industry is in many ways ideal; principally because the towns within Britain were all almost equally feasible as potential adopters. Even though the cost of fuel did differ from one area to another depending on the distance of a town from access to coal supplies, this differential input cost was counterbalanced by the fact that the main by-product, coke, could be sold at higher prices in areas in which coal itself was more expensive (Falkus 1967, p. 500). Further, even though the financial returns were potentially higher in towns with large populations and with a larger number of commercial and manufacturing enterprises, gas companies could be – and, indeed, gradually were – established in relatively small places at quite early dates. The scale economies appear not entirely to have excluded small towns from the ranks of potential adopters of the innovation, and the spread of gasworks extended throughout a wide range of town sizes within two decades after the foundation of the first company. The whole population of towns was therefore equally 'at risk'.

Falkus' conclusion that gasworks diffused regularly down the size hierarchy of British towns during the first half of the century adds considerable substance to the emphasis placed on the size of towns in the study of building societies and telephone exchanges. Is it possible, in studying the data on gasworks, to arrive at some more precise evaluation of the respective importance of hierarchical diffusion and neighbourhood spread for this particular innovation? The details of the earliest dates of establishment of gasworks in towns in England and Wales

Table 5.8 *Hierarchical diffusion of gasworks in English towns, 1812–40*

Rank of towns	*Maximum number of towns*	*Number of towns with gasworks by the end of:* 1820	1829	1840	*Percentage of places with gasworks by the end of:* 1820	1829	1840
1– 10	10	8	10	10	80·0	100·0	100·0
11– 30	20	9	20	20	45·0	100·0	100·0
31– 60	30	12	25	25	40·0	83·3	83·3
61–100	40	3	23	32	7·5	57·5	80·0
101–150	50	1	12	32	2·0	24·0	64·0
151–210	60	1	8	36	1·7	13·3	60·0
211–280	70	0	6	39	0·0	8·6	55·7
281–360	80	1	3	35	1·6	3·8	43·7
361–450	90	–	0	24	–	0·0	30·4
'Towns' < 2,500		0	4	33			
Places < 2,500		0	0	26			

Notes

1. Towns are ranked by their populations in 1821, 1831 and 1841 respectively.
2. The two final rows – '"Towns" < 2,500' and 'Places < 2,500' – are places in which gasworks were built before they reached the threshold of 2,500 population. The first are places which did exceed that threshold at some date before 1911; the second are places which did not exceed the threshold before that date.
3. The figures in the three final columns are calculated as percentages of the number of towns in each category of ranks.
4. Approximations to the population sizes corresponding to town ranks can be taken from fig. 2·3 (p. 30). They will be approximate because of the exclusion of Welsh towns from this table.

Source: Falkus, private communication.

were very kindly made available by M. E. Falkus and it was hoped to be able to suggest answers to this question. Table 5.8 shows the progressive spread of gasworks at three dates for towns within England.[1] The three dates have been selected in the light of the cyclical pattern of the construction of gasworks shown in fig. 5.13: 1820 marks off a period of 'early adopters';[2] 1829 and 1840 include the construction booms in the middle years of the 1820s and 1830s respectively. The most striking feature of the pattern is undoubtedly the close connection between the date of adoption and the size of town. By the end of

[1] Welsh towns have been excluded from consideration since the data appear less reliable in their case.

[2] Indeed by as early as 1822 no fewer than seven gas plants had been established in London (*Parliamentary Papers* 1823, V, 308-41).

1820, all but two of the ten largest English towns, and all but seven of the largest twenty towns had gasworks. Of the exceptions, the great majority had constructed gasworks very soon after 1820: the two towns in the top ten – Plymouth and Portsmouth – by 1823 and 1821 respectively; and of the five towns in the next ten, four had gasworks by 1821 – Norwich, Hull, Leicester and Huddersfield, and the fifth – Sunderland – had a gasworks by 1824. By the end of 1829, all of the thirty largest places had companies and all but five of the top sixty towns had adopted. As Falkus suggests, there was indeed a very close correspondence between size and date of adoption.

The tail of the distribution is, however, fairly distended. At the same time that large places were adopting, gasworks were also being built in very much smaller places. Even before 1820, for example, a town as small as Helston in Cornwall was lit by gas. Other examples of relatively small towns which were

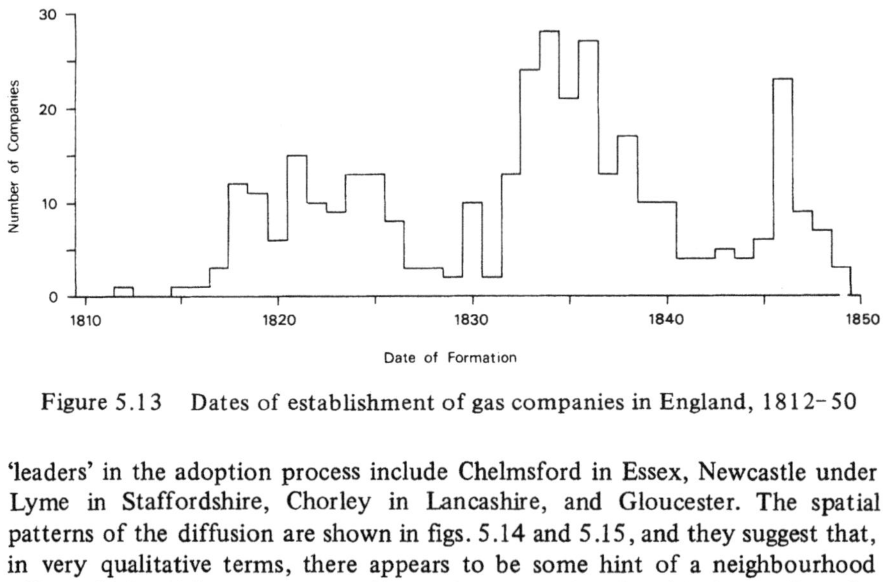

Figure 5.13 Dates of establishment of gas companies in England, 1812–50

'leaders' in the adoption process include Chelmsford in Essex, Newcastle under Lyme in Staffordshire, Chorley in Lancashire, and Gloucester. The spatial patterns of the diffusion are shown in figs. 5.14 and 5.15, and they suggest that, in very qualitative terms, there appears to be some hint of a neighbourhood effect which might go some small way in accounting for this long tail to the distribution of adopters. One marked feature is that gasworks tend to be established early in the county towns of the regions within England, whether or not such places were large. Gloucester was certainly not among the largest of towns; neither were Hereford, Lancaster, Chelmsford, or Reading. All, in fact, fell below the top fifty towns in terms of population size and yet adopted gasworks at relatively early dates. Such 'precocious' adoption might reflect their importance with respect to flows of communication and contact, deriving from their long-standing role as regional centres before the advent of industrialization.

The maps of diffusion also hint at a possible pattern of regional development of gasworks in which towns within a particular area tend to be developed within

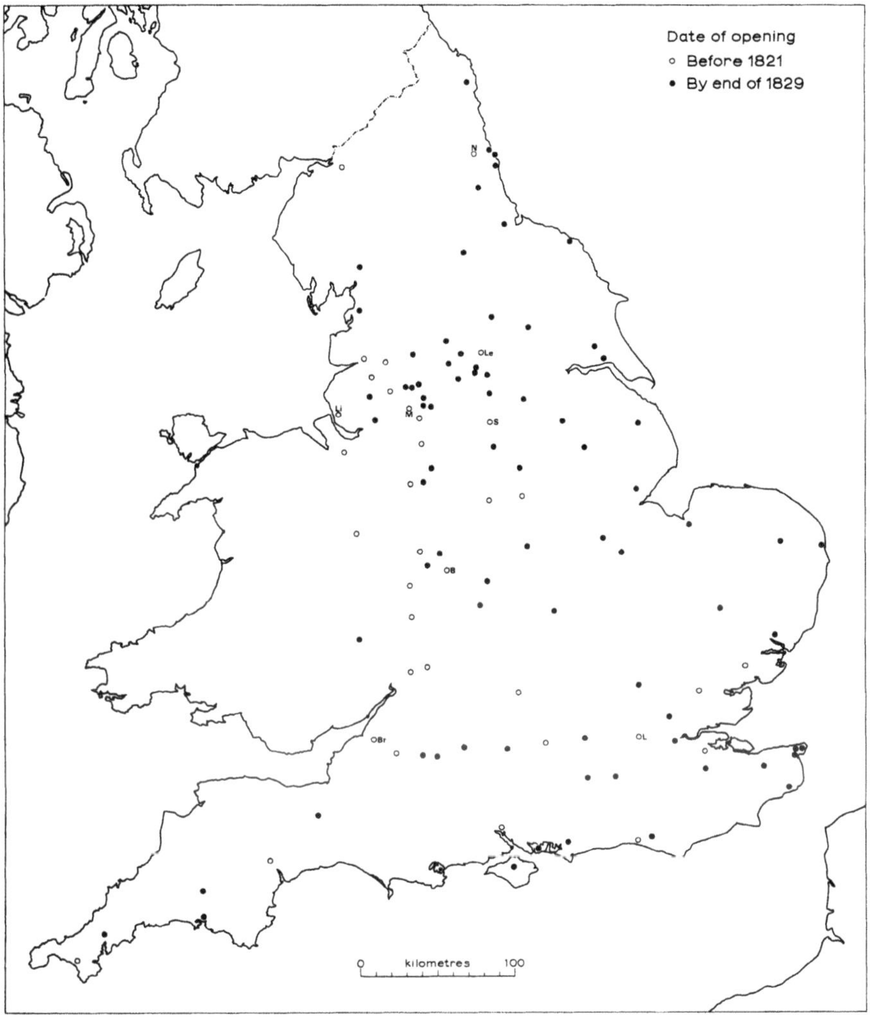

Figure 5.14 Gasworks in England: spatial diffusion 1820–9

a relatively short span of time. Lincolnshire, for example, despite the fact that it contained relatively few large towns, was developed with gasworks relatively early. As in the diffusion of building societies, towns close to London tended to be developed earlier than their size alone might have led one to expect; and this was particularly true amongst the towns in Kent and along the coast of Sussex. The towns of coastal South Devon had gasworks within relatively few years of each other in the decade of the 1830s. The same, again, is true of Dorset towns: Bridport was lit in 1832; Dorchester and Poole were both lit in the following

Figure 5.15 Gasworks in England: spatial diffusion 1829–40

year; Lyme Regis in 1835; and Blandford, Shaftesbury and Sherborne all in 1836. Some part of this apparent tendency for neighbourhood diffusion must be attributed to the contractor system by which many of the gasworks were constructed; plant being built either on local initiative and under the superintendence of a specialist gas engineer called in from outside, or through the activities of speculators who would build plant and subsequently sell it to a company of shareholders. Falkus (1967, pp. 505-9) gives instances of the activities of such promoter-contractors, many of whom appear to have operated on a regional basis. The Malam brothers, for example, who were responsible for over fifty gasworks before 1846, concentrated their attention primarily within distinct regions of the county: one in Lancashire and the West Riding of Yorkshire; another in Cumberland and the North and East Ridings; and the third in East Anglia.

On the basis of such evidence, it is possible to suggest, in a largely qualitative fashion, that the two principles of hierarchical and neighbourhood diffusion can be seen to have operated in the spread of gasworks in the first half of the century. A more quantitative approach, which might have isolated the respective weightings of the two principles however has been frustrated by the imperfections of the basic data and by the temporal progression of the construction of gasworks. On the first point, it is evident that the dates of first establishment of gasworks are not as reliable as would be required for a more precise analysis. Most of the larger towns were lit by companies which were large enough to justify their incorporation as limited liability companies and the dates of the appropriate Acts provide a good indication of the date of their first operations, but in the smaller places many of the undertakings were not incorporated and, where they were eventually incorporated, the dates of the Acts given in the gas directories usually do not indicate the date at which the original firm began operations. Falkus drew on the records of a firm which supplied castings to the gas companies and, on the basis of a sample drawn from this source, was able to suggest the dates of construction of many gas works in the widely scattered smaller towns throughout the country. The sphere of operations of this firm was, however concentrated in northern and eastern England and there is certainly a number of towns for which the earliest dates which can be established are not those of the first opening of gasworks and the lighting of the town. For example, one of the most obvious anomalies in the data presented here is the town of West Bromwich which was the largest town in 1841 shown as not having been lit by 1840. Indeed, gas directories[1] suggest that a company began operations in West Bromwich as late as 1880. Yet clearly there was a gasworks well before this date. Lewis' *Directory* (S. Lewis 1844, volume 1, p. 389) noted of West Bromwich in 1844 that: 'Very large gas-works have been established by a company of proprietors, from which part of Birmingham, seven miles distant, is lighted, and from which the different towns of Wednesbury, Dudley, Bilston, Darlaston, Tipton, and Great Bridge, are also supplied with gas.' Other such anomalous entries in the data which have been used must doubtless be attributed to similar difficulties in establishing the date of the very first company to start operations in a town – as against the dates of subsequent companies or of extension to existing companies.

The second fact which militated against a more precise analysis of the data on gasworks was the time-curve of construction. The fact that companies were

[1] For example, *Gas . . . companies' directory* 1881.

developed in so markedly cyclical a fashion means that equal time increments had very different significance in the cumulative curve of the adoption of gasworks. A difference of five years between dates of openings would be of very different significance during a period of boom in construction than in a period of slump when few plants were built. If a form of regression were to be used in the study of the diffusion of gas lighting, the dependent variable – date of construction – would therefore need to be weighted in some way to take account of this difference and the weighting would be of a complex form, not constant over the entire length of time from 1812 to 1840. The problem is similar to that in Berry's (1972, pp. 343-4) study of the diffusion of television stations in America, where he notes that the two-year freeze on construction during the Korean war had the effect of causing a hiatus in the logistic curve of openings during the years 1950 to 1952.[1] Places which would have been expected to open stations during these years, on the basis of their population size, postponed the openings until the post-war period during which time there consequently was a much wider scatter of sizes of places in which television stations were opened. The same effect must have operated in the case of gasworks because of the correlation of construction with the swings of the general trade cycle.

Despite the absence of a more quantitative analysis of the spread of gasworks, however, the descriptive approach adopted here and in Falkus' analysis, do suggest the importance of town size and, to a lesser extent, of localized spread in the diffusion of gas lighting.

CONCLUSION

The diffusion of gas works at the start of the century, the spread of building societies in the middle years of the century and the development of the telephone in the final quarter of the century each illustrate the operation of the twin processes of hierarchical diffusion down the ranks of the urban size array and the localized 'neighbourhood' spread out from early adopters to smaller towns within their ambit. The relative importance of the hierarchical and neighbourhood effects in each case depends principally on the characteristics of the innovation itself, but, whatever their respective weighting, it would appear that both processes have to be taken into account in any attempt to simulate or to understand the most probable course of the diffusion of entrepreneurial innovations.

The particular innovations which have been looked at in some detail here may in themselves have given rise to relatively little direct employment. The significance of the nature of their spread, however, is that it suggests the validity of using a rather general model of diffusion which incorporates hierarchical and

[1] The same tendency was produced by the effects of the Second World War.

neighbourhood principles. Whether or not the particular innovations directly 'caused' population growth in towns, it would certainly seem probable that their more indirect effects were not inconsiderable. The fact that a town adopted gas lighting, or developed telephone exchanges, or that it had flourishing building societies is also in itself some indication of its 'vitality' and 'progressiveness' as a town – qualities held in some esteem in the expansive, quantity-conscious Victorian age. It is not without significance that many contemporaries themselves stressed just these very qualities as ones which attracted immigrants to the towns. H. L. Smith in his well-known discussion of migration in Booth's *Life and labour of the people in London*, considered that the lure of the town was:

> . . . the contagion of numbers, the sense of something going on, the theatres and the music halls, the brightly lighted streets and busy crowds – all, in short, that makes the difference between the Mile End Fair on a Saturday night, and a dark and muddy country lane, with no glimmer of gas and with nothing to do. Who could wonder that men are drawn into such a vortex, even were the penalty heavier than it is?

And if the local 'metropolis' was not only gas-lit, telephone-served and offered the lure of building societies, but also had the hundred-and-one other urban 'improvements' of the age, no doubt the pull of the vortex, both for those within and those without, was stronger still. The nature of the diffusion of entrepreneurial innovations may well serve as a guideline to the manner by which the general growth of nineteenth-century towns took place.

CHAPTER 6

Simulating urban growth

But who Pretender is, or who is King,
God bless us all – that's quite another thing.

JOHN BYROM

The evidence of the three general indicators which have been studied suggests that the diffusion of entrepreneurial innovations through a set of nineteenth-century cities may well be characterized by a simultaneous process of hierarchical space-jumping from larger to smaller cities and of localized spread outwards from regional centres. The particular balance between these two components appears to be determined by the characteristics of the particular innovation being studied – primarily by the importance to it of scale economies and the threshold market size which it requires. Where a large market is called for – because of the need to recoup heavy initial capital outlay, or to minimize the element of risk, or because of marked scale economies – the hierarchical effect may be expected to be commensurately more important in determining the course of the diffusion process. As technical changes alter the costs of installation or operation of a given innovation, it may be that the innovation subsequently filters down to places of much smaller size and that the spread effects become more important in its diffusion – as indeed appears to have been the case with the gas industry. Alternatively, however, it may be that technical improvements may well *increase* the optimal scale of production for an innovation so that, over time, it is less likely to diffuse to smaller places (Pederson 1970, pp. 219-20). Where an innovation is not greatly affected by size constraints – as in the case of building societies – the spread, or neighbourhood, effect may be apparent at the very outset of the diffusion, but it would nevertheless appear that hierarchical diffusion down the ranks of town sizes is rarely, if ever, absent. Even when neighbourhood spread is important, innovations still appear to be diffused to new regions, which are distant from the originating centre, initially through their large towns or regional capitals which act as the organizing poles of economic and social activities within the region.

Underlying the attempts to provide validation of this twin process of diffusion in the nineteenth century has been the assumption that the adoption of innovations is a cause, as well as an indicator, of economic vitality in towns and is therefore closely tied to their population growth. We are now therefore in a position to attempt the development of a model to simulate the growth of towns on the basis of the radically simplified assumption that urban growth is 'caused' by the diffusion of entrepreneurial innovations. Some of these innovations in the real world will, of course, be of a rather specialized technical nature and only those few places which either pursue or which come to pursue the appropriate economic activity will be potential adopters; improvements in mine-ventilation techniques would be of little interest to the inhabitants of Bournemouth. Others, however, will be more universal innovations which would have potential benefits to offer to the whole range of places. We can, however, think of the whole set of innovations as averaging out into a 'composite' general innovation which will produce growth in the populations of those places which adopt since the innovation will have both direct effects on the growth of employment as well as more indirect effects by making the towns more efficient or more attractive places both socially and economically. Both in terms of retaining people born within the town and attracting migrants from without, the adoption of the composite innovation will therefore have the effect of increasing the town's population.

Our objective in simulating the diffusion of such a 'composite' innovation will be to see to what extent this urban growth can be produced in a way compatible with the two aspects of the growth of towns suggested by the empirical data; a scatter of growth rates whose variance declines with increasing town size and an average growth rate which can vary between one in which size and growth are positively related to one in which the two are not related. Let us assume that we have a set of towns with given population sizes and given locations and that certain of them have, and others have not, adopted the 'composite' innovation. Those which have adopted will send out messages – or stimuli – about the innovation. For places which have not adopted, these stimuli will represent information about the innovation and, if a non-adopter town receives a sufficient number of such stimuli, it will adopt and in turn will itself begin to emit stimuli. For places which have adopted, the receipt of stimuli will represent agitation in the local economy and will lead to further employment and an increase in the town's population. From our discussion of the nature of innovation diffusion, it can be assumed that the spread of the innovation can be represented by a form of gravity equation, which will combine both hierarchical and neighbourhood effects. For a town, i, with population P_i, the probability of its receiving a stimulus will depend on its population size, its distance from towns which have adopted and which therefore send out stimuli, and the number of stimuli which are emitted by those towns. We therefore have an equation of the following sort which describes the probability of receiving stimuli:

$$\mathrm{Pr}_i = k\left[\sum_{j=1}^{n}(P_i^a \, . \, d_{ij}^{-b} \, . \, S_j)\right]; \qquad [6.1]$$

where the summation is over all of the set of j-places ($j = 1 \ldots n; j \neq i$) which have already adopted; S_j is the number of stimuli sent out by the jth adopter; and k is a weight which converts the sum of the expressions for all of the places to 1·0, so that one is dealing with probabilities. For the total set of towns, we therefore have a given number of stimuli being emitted by those places which have adopted the innovation and, for each town, whether or not it has adopted, we have a probability of it receiving a stimulus. The stimuli can thus be allocated amongst the set of towns to cause them either to adopt the innovation if they have not already done so, or to grow in size if they have already adopted. To incorporate into the model the chance elements so characteristic of the real world, this allocation of stimuli can be made on a probabilistic basis. This would involve giving sequential numbers to each of the towns – the range of numbers being dependent upon the town's probability of being 'hit' by a stimulus – and then drawing a random number for each stimulus which is to be allocated. This is a type of Monte-Carlo method which, like Hägerstrand's simulation of spatial diffusion, neatly combines the deterministic and chance elements of the real world: the first in terms of the structure of equation [6.1] which determines the set of probabilities; and the second in terms of the use of random numbers.

To illustrate the procedure with a simple example, let us take a set of five places, only two of which have adopted the innovation and which send out five and three stimuli respectively. The distances and sizes of these towns will be as in table 6.1. If we let the a and b exponents in equation [6.1] each be 1·0, the calculation of the probability of place 2 being hit would be:

$$\mathrm{Pr}_2 = k \left(\frac{600}{10} \times 3\right) + \left(\frac{600}{12} \times 5\right) = k \times 430.$$

The value of k, which converts these sums so that they add to 1·0, will be 1/826 and this produces the column of probabilities given in table 6.1. Since we now wish to distribute the eight stimuli amongst the set of towns, these probabilities are converted to sequential numbers, as in the final column of table 6.1, and eight random numbers are drawn to allocate the stimuli. Supposing that the random numbers were 69, 60, 11, 74, 15, 4, 21 and 97, this would mean that place 1 would receive two stimuli, place 2 would receive three, place 3 would receive two, place 4 would receive one and place five would receive none.[1]

[1] It can be seen that equation [6.1], by summing over all towns other than the adopter town itself (i.e. $j \neq i$), sidesteps the question of the number of stimuli a town sends to itself. This broaches the unresolved problem in the literature of social physics of how best one might calculate the 'self-potential' of a town. However the fact that in the development of the simulation model, stimuli are distributed in an aggregate fashion and also that some units of population growth are given to a town once it has adopted, whether or not it receives further stimuli, justify our ignoring the self-potential of adopter towns. The logic of the model is that growth – or agitation to the local economy – will occur principally through the interaction *between* places rather than *within* a place.

Having allocated the stimuli, two things have then to be decided: at what point will adoption occur for a town which has not adopted; and by how much will the population of a town which has adopted grow during the time period represented by this sending-out of stimuli. The first decision would represent the level of resistance which would exist in a town before it decided to adopt and, in the real world, this would depend upon the type of innovation involved, its qualities of perceived risk and its expected costs and returns. Such resistance levels may well differ between towns, with one place being more adventurous or, indeed, more suited to the particular innovation. But in our example we could say that the resistance level was the same for all places and that a place would need to have received three or more stimuli before it would adopt. Place 2 would thus adopt during the time period; place 4 would not adopt, but, having received one stimulus, its resistance would be lowered so that it would require only a further two stimuli in subsequent time periods to cause it to adopt; place 5

Table 6.1 *A simple example of the diffusion model*

Place	*Units of population*	*Units of stimulus emitted*	*Units of distance to place:* 1	2	3	4	5	*Sum of contacts*	*Probability of being 'hit'*	*Sequential numbers allocated*
1	400	3	–	10	20	15	5	100	0·12	1– 12
2	600	–	10	–	12	8	8	430	0·52	13– 64
3	1000	5	20	12	–	5	20	150	0·18	65– 82
4	100	–	15	8	5	–	15	120	0·15	83– 97
5	40	–	5	8	20	15	–	26	0·03	98–100

would remain unaffected. Second, for places 1 and 3 which had already adopted, we would need to decide by how much the receipt of their two stimuli would cause their populations to grow. It could be that each stimulus would cause incremental growth of some fixed amount or alternatively of some fixed percentage of the existing town population.

Whatever were the eventual rules which were decided, this outline provides the skeleton of the simulation model to which the details of the operating rules would add flesh. We have a set of towns, some of which have adopted the innovation, and the stimuli sent out from them are allocated amongst the towns, causing certain of those which have not adopted to adopt and producing growth amongst those which have already adopted. This gives one cycle of the model at the end of which there will be a new set of populations, a new set of adopters, and a fresh and larger reservoir of stimuli to be distributed in the next time period. The nature of the model is very simple and, being simple, is far removed from reality. In essence, it is modelling a form of composite innovation which generates factories, employment and general economic vitality. In reality there would be a large variety of innovations, each with different characteristics and

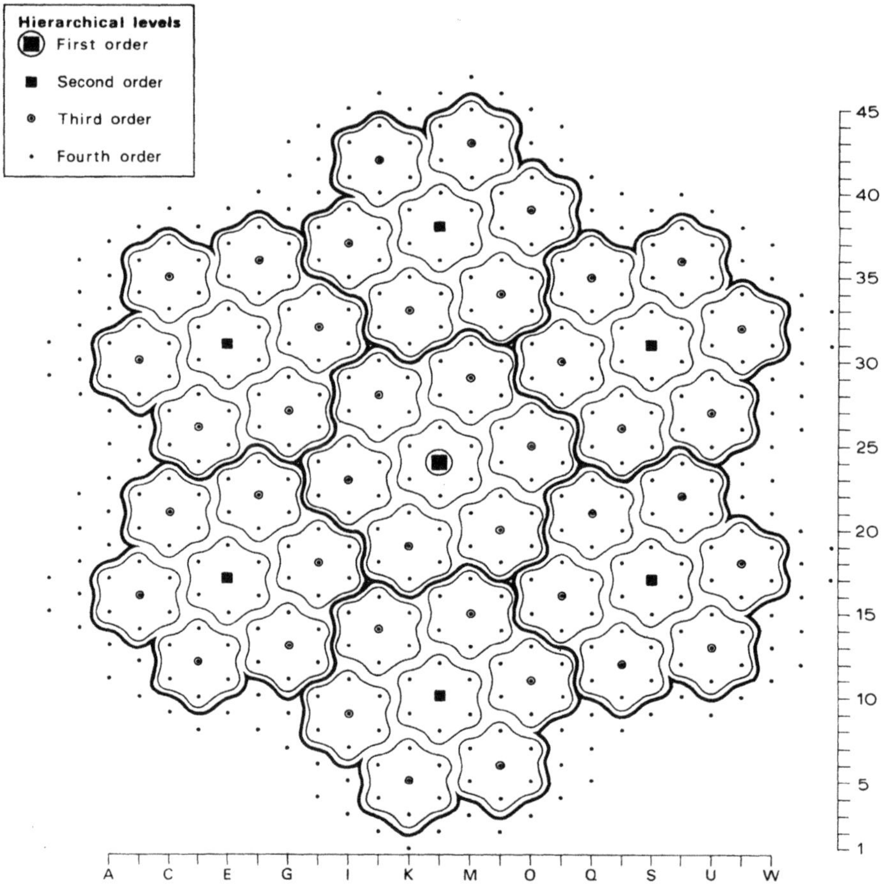

Figure 6.1 An idealized central-place system of towns. An arrangement according to Christaller's K = 7 network

each having effect on the growth of population of either all or some of the towns. For any one of the innovations, the effect on growth would almost certainly decline over time; both in terms of the fact that economic returns may be lower for later adopters and that, within a given town, one innovation would eventually be superseded by others. But, in effect, the simulation model pools together the effects of all of these innovations by assuming a continuing stream of inventiveness and innovation over time. It treats the stream of innovations in an aggregate fashion rather than first diffusing one innovation and, while this first one is still spreading eventually to reach a saturation point, beginning the diffusion of a second, a third, and so forth.

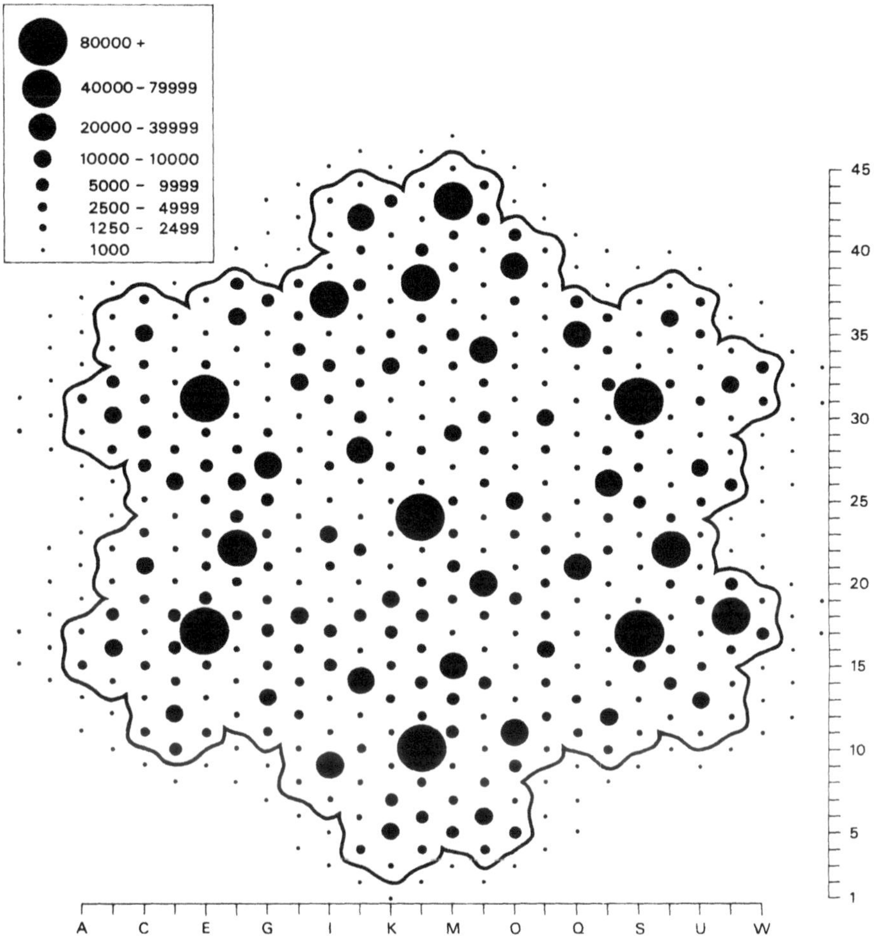

Figure 6.2 Population sizes in the idealized urban system. Populations form a rank-size array and are allocated randomly to cities within each level of the hierarchy shown in fig. 6.1

To examine the nature of the growth processes which this model produces, we shall consider a set of towns distributed in an idealized landscape and arranged according to Christaller's central-place arrangement with $K = 7$.[1] The distribution of towns is shown in fig. 6.1 and it can be seen that there are four

[1] This arrangement was chosen since it is the only one of Christaller's three basic geometrical arrangements in which the boundaries of the market areas of each order of central place coincide. It is therefore possible to draw a 'self-contained' urban system with an overall boundary which includes the market areas of all places.

hierarchical orders of town: one first-order place; six second-order places; forty-two third-order places; and 294 fourth-order places. It is amongst this set of towns that the mythical innovation will be diffused. In addition to the towns within the central-place arrangement, there are also 112 places which have been added around the perimeter of the area. These have been included to add what Hägerstrand calls 'border bounce' into the 'system' and so help to alleviate (but not to solve) the difficulty of dealing with a bounded set of places.[1] Each of the 343 towns was then allocated population sizes in such a way that the overall size distribution conformed to a rank-size curve: the largest place having 500,000 population; the second largest having 500,000/2; the 343rd largest having 500,000/343, or 1,458 population. These populations were randomly allocated within each of the four hierarchical orders of towns, so that the first-order town was given 500,000 population, the six second-order towns were randomly assigned populations of between 250,000 and 71,279, the forty-two third-order towns were randomly assigned populations between 62,350 and 10,204, and the fourth-order towns were randomly assigned populations of from 10,000 to 1,458. The remaining 112 towns which lie outside the boundary of the area were each allocated a population of 1,000. The resulting distribution of towns, shown in groups of town sizes, is given in fig. 6.2. This initial set of towns is obviously only one of an infinite variety of locations and sizes which could have been chosen, but it attempts to approximate the empirical evidence on urban size distributions and also the theoretical notion that the distance between large towns is greater than that between small. Since the form of the basic equation within the model (equation [6.1]) relies both on spacing and town size, however, the results will obviously be influenced by the particular distribution from which one starts, and the pattern of fig. 6.2 needs therefore to be borne in mind in the interpretation of the subsequent results.

We are now in a position to establish the operating rules of the simulation. From the derivation of the model it is evident that there is a variety of parameters and attributes which can be altered. First are the parameters of equation [6.1]; the distance exponent could be raised or lowered so as to give greater or less effect to the neighbourhood process of spread and the weighting of population could equally be altered. Second, there are numerous ways in which one could determine the number of stimuli which would be emitted by an adopter town; the most realistic might be to make the number depend upon the population of the town in some way. Third, there is a variety of ways in which

1 See Hägerstrand (1967b, p. 18). The boundary problem is a recurring one in spatial studies. Here the difficulty can be seen by comparing the potential set of places from which a central and a peripheral place can receive stimuli: a central place has towns within all quarters of the compass; whereas a place lying close to the southern boundary only has towns lying in the quarter to its north. The 112 border towns are therefore added so that they can 'catch' any stimuli sent outside the 'system' and, if they themselves adopt, can subsequently send stimuli to the border towns within the system. We shall therefore not be interested in the growth rates of the border towns themselves.

population could be increased for each stimulus received by an adopter town; it could be raised by a given number of people or by a fraction of the existing population. Fourth, the levels of resistance which determine the time at which a town adopts could be altered; a higher or lower figure could be chosen or the figure could be made to vary from one town to another in some way. Each of these decisions will have effect upon the operation of the model and each could be interpreted in terms of the changes which it represents in the real world – a topic to which we shall return below.

MODEL 1

The operating rules for the first version of the simulation – Model 1 – were established as follows:

1. At the outset, time t_0, only the largest town has adopted the innovation.
2. Towns adopt once they have received three stimuli from adopter towns. Having adopted, in subsequent time periods their populations grow as specified in rule 5 and they emit a certain number of stimuli as in rule 3.
3. Adopter towns send out S stimuli in each time period subsequent to their adoption:

   $$S_j = \sqrt[4]{P_j};$$

   where P_j is the population of the jth adopter.
4. The probability of town i being 'hit' by a stimulus is given by the following:

   $$k \sum_{j=1}^{n} P_i . d_{ij}^{-1} . S_j \quad (j = 1, \ldots n, j \neq i).$$

5. Once a place has adopted, in subsequent time periods its population grows in the following way:
   (i) every fresh stimulus which it receives generates S additional population ($S_j = \sqrt[4]{P_j}$) in the time period in which the stimulus is received and $\frac{1}{2}S_j$ population in the subsequent period.
   (ii) once it has adopted, a town generates its own further growth by the addition of S_j population in every subsequent period, irrespective of whether or not the place receives stimuli.

   For any given period, population growth will therefore be given by the following:

   $$N_{j.t} . S_{j.t} + N_{j.t-1} . S_{j.t}/2 + S_{j.t};$$

where $N_{j.t}$ is the number of stimuli received by place j in period t and $N_{j.t-1}$ is the number of stimuli received in the preceding period.

These particular rules for population growth and the generation of stimuli need some brief elaboration. Rule 5(ii) argues that once a town has adopted, it will begin to generate a certain, small, amount of population growth of its own. This is tantamount to adding a 'self-potential' of a place to the gravity-based equation of rule 4[1] which would represent the agitation caused to the urban economy by the fact of its having adopted. This adds a deterministic element to the model since it means that, once having adopted, a place will grow, even if only by a very small amount. The main determinant of growth, however, will still be the stimuli that are received from the 'system' at large and rule 5(ii) argues that a place will grow primarily if, through its contacts with other innovating centres, it continues to take part in the general ferment of the innovative sequence. The effect of each stimulus will have an immediate impact on population growth in the period at which the stimulus was received and also a further, although modified, effect in the next time period. Rule 3 determines that the number of stimuli sent out in each time period is a weighted function of the population size of the town which has adopted; large towns will send a larger absolute number of stimuli, but a smaller proportionate number, than will small towns.

This version of the basic simulation was run over a number of time periods for the set of towns shown in fig. 6.2 and the growth rates of the urban populations were calculated at the end of each period using the compound growth-rate formula given in chapter 3.[2] The pattern of growth rates by the end of the tenth period is shown in fig. 6.3. Only those places which had adopted, and whose population had therefore changed, are shown.[3] The pattern of growth is interesting since, while there is a decreasing scatter of variance at larger sizes of towns, the upward curve of the average growth is far steeper than was ever found in the data of the nineteenth century. The source of this discrepancy is very readily apparent in the nature of the rules of Model 1, since the combination of rules 3 and 5 makes the effect of innovations on urban growth doubly multiplicative; rule 3 making the number of stimuli a function of town size and rule 5 making the increments of additional population also a function of size. If we were considering only a single innovation it would be reasonable to argue that, once a town had adopted, its population would subsequently grow by some fraction of its existing size and thereby simulate the multiplier effect of innovation diffusion, but in our model we are dealing with a whole stream of innovations and the multiplier effect is already incorporated into the model by the increasing numbers of stimuli which a place is likely to receive once it has adopted. It would therefore be more realistic to argue that, given the increasing flow of stimuli in the model, each individual stimulus received by a town would

[1] See above, p. 193.

[2] Equation [3-4], see above, p. 71.

[3] Of the total of 343 places, sixty-nine had not adopted by period 10 and thus had growth rates of zero.

represent a fixed increment of population. This increment would then represent, say, the addition of a factory or of a unit of tertiary employment which would be of the same size irrespective of the size of town to which it was added. The only difference between the likely growth of a large and a small place would then be that the large place would be more likely to attract a greater number of such incremental units by receiving more stimuli through the structure of the equation in rule 4. The multiplier effect would be contained in the tendency for stimuli to increase at something approaching a geometric ratio.

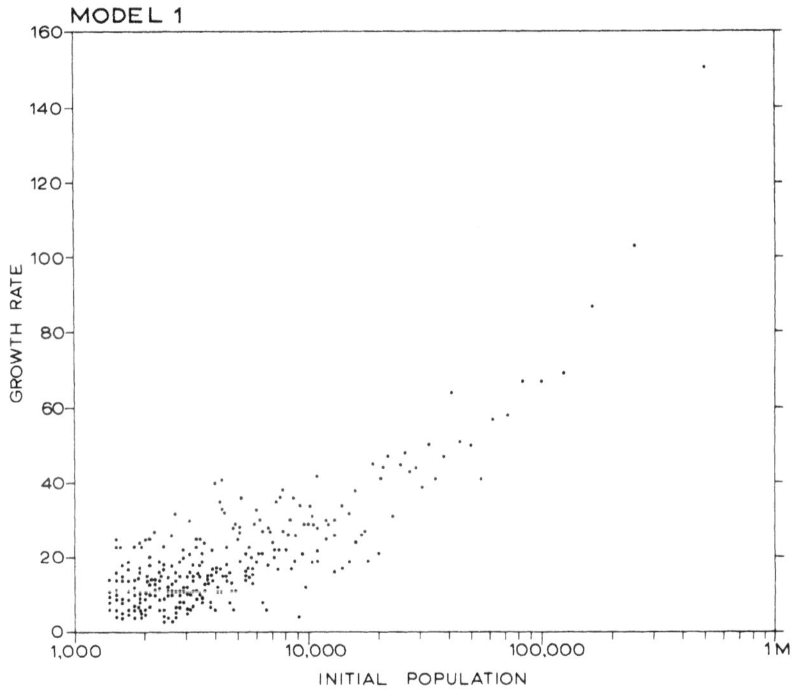

Figure 6.3 Urban growth rates: Model 1 at generation 10

MODEL 2

Model 2 therefore has the following rules:

1-4. As in Model 1.

5. Once a place has adopted, in subsequent time periods its population grows in the following ways:
   (i) every fresh stimulus which it receives generates 100 units of additional population in the time period in which the stimulus is received and 50 units in the subsequent period.

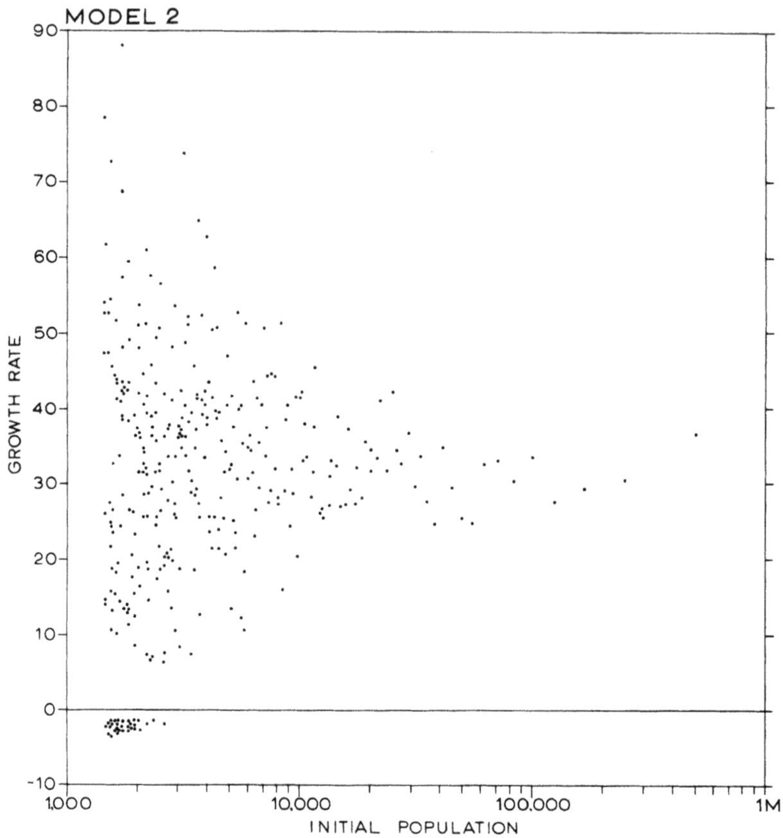

Figure 6.4 Urban growth rates: Model 2 at generation 10

(ii) once it has adopted, a town generates its own further growth by the addition of 100 units of population in each subsequent period, irrespective of whether or not the place receives stimuli.

For any given period, population growth will therefore be given by the following:

$$100(N_{j.\,t}) + 50(N_{j.\,t-1}) + 100.$$

Model 2 also differs from Model 1 in that it allows the population of a town to decline if the town does not receive sufficient stimuli. Since the total number of stimuli available is small in the first few periods of the simulation this decline in population only begins to operate after a certain number of cycles of the model. This additional rule would read:

Table 6.2 *Distribution of dates at which places adopt: Model 2*

Period	*Size group*							*Total*
	1	*2*	*3*	*4*	*5*	*6*	*7*	
0	1							1
1	1							1
2	4		1					5
3		4	3	1	1	1		10
4		2	8	3	2	2		17
5			1	15	10	4	5	35
6				6	17	20	4	47
7					15	31	16	62
8					3	23	37	63
9					2	14	28	44
10						4	24	28
11						1	13	14
12							9	9
13							3	3
14							3	3
15							1	1
After 15								0
								343

Note: Size groups are as follows:
1 80,000+
2 40,000–79,999
3 20,000–39,999
4 10,000–19,999
5 5,000– 9,999
6 2,500– 4,999
7 1,250– 2,499

6. If, in the periods after the second cycle of the model, a place fails to receive any stimuli over two consecutive periods, its size will decline by S population ($S_i = \sqrt[4]{P_i}$).

This rule draws some distinction between those places which, irrespective of whether or not they have adopted, continue to receive stimuli and those places which do not. It argues that, whether or not a place has adopted, if it does not take part in the general innovative ferment its population will decline.

Running Model 2 produces a markedly different pattern of growth from that of Model 1. At the end of the tenth period (fig. 6.4), there is only a relatively small number of places which have not adopted. The populations of some of these places have declined by an amount depending on the number of stimuli received since some of the non-adopters may still have received some, but less than three, stimuli which would have had the effect of slowing their rate of decline. The growth rates of the whole set of places shows a scatter which is remarkably, and encouragingly, similar to that of the real world. The largest place, with its initial advantage of having begun the whole growth sequence through having had the innovation from the outset, has a somewhat higher growth rate than expected for its population size, but, overall, the curve of average growth shows a roughly constant average irrespective of size and the most striking feature of the scatter is the way in which the variance decreases with increasing town size. The way in which this declining variance is produced can be seen from the sequence of adoption dates and its relation both to the size and the location of the towns. Table 6.2 shows the date of adoption for groups of towns of different size. The distribution bears a close relationship to the size-related diffusion processes of nineteenth-century entrepreneurial innovations. All of the towns have adopted by the fifteenth period, but the average date of adoption becomes successively later for groups of progressively smaller towns: all of the largest group has adopted by period two; all of the second largest group by period four; all of the fifth largest group by period nine; and so forth. There is, however, an increasing variance in the date of adoption of groups of smaller-sized towns: for example, in the third largest group one town adopts as early as period two, while the latest adopter is in period five; for the very smallest group this range extends from periods five to fifteen. This again was seen to be a feature of the process of diffusion in real-world data; certain small towns adopt precociously early even though the average date gets later for groups of smaller towns.

There are two reasons which account for this variance in the date of adoption (and for the corresponding increase in the variance in growth rates for smaller places).[1] First is the chance or stochastic element in the simulation model which means that, even though the probability of a small place receiving a sufficient number of stimuli for it to adopt may be small, there is still a probability that it will adopt and grow and so will enhance the future likelihood of its receiving stimuli. Second, however, is the influence of the location of the towns which is incorporated into the structure of the model through the neighbourhood effect of distance friction. This second effect can clearly be seen by looking at some specific instances of the adoption sequence which are shown in fig. 6.5. For example, place J.28,[2] with a population in the third-largest group of sizes, adopted as early as the second period, no doubt as a result of its relative proximity to the largest town. This had consequences, subsequently, for the small places in the vicinity of J.28 since they then lay in the vicinity of a place

[1] With so large a number of places, it is easier to examine the results of the model in terms of date of adoption. This is closely related to growth rate in the early generations of the model, but less close at later time periods when the advantage of early adopters tends to be nullified by the rapid growth in the number of stimuli which causes the different variants of the simulation to converge to a similar end-product set of growth rates.

[2] Towns are referred to by the co-ordinates shown in fig. 6.2. The largest, central, town has a co-ordinate reference of L.24.

emitting stimuli from a very early date and thus would have a higher-than-average expectancy of being 'hit' by stimuli. One of them, place J.30, adopted as early as period 4 and had received a total of thirty-nine stimuli by period ten and had a growth rate of 51 per cent – well above the average date of adoption. number of stimuli and growth rate of places of comparable population size. Similarly, the effects of proximity to the central largest town can be seen on the growth of the towns immediately surrounding it. Despite the fact that the random allocation of initial populations to these places happened to give them

Table 6.3 *Diffusion and growth in small towns adjacent to the first-order town: Model 2*

Place	*Initial population*	*Period of adoption*	*Number of stimuli by period 10*	*Growth rate by period 10 (%)*
L.26	1,736	5	15	88
M.25	3,205	5	23	74
M.23	3,125	8	9	35
L.22	1,565	6	11	73
K.23	1,623	7	7	52
K.25	1,587	9	4	26
M.27	1,930	8	6	39
N.24	1,462	10	3	14
M.21	6,493	4	24	44
K.21	1,511	11	2	−2
J.24	2,415	8	5	25
K.27	4,000	8	12	34

Notes

1. Places are referred to by co-ordinate references as given in fig. 6.2. They are arranged in the table in clockwise order around the central town, L.24. The first six places are closer to L.24 than are the second six places.
2. Dates of adoption should be compared with those of other towns of comparable size given in table 6.2.

very small populations in comparison with the average fourth-order town, their dates of adoption tended to be earlier than the average for their respective sizes. Indeed, if one looks at the fortunes of the twelve fourth-order towns which lie closest to the largest town (table 6.3) the nature of the diffusion process in the simulation model can be seen rather clearly. There is considerable variability amongst the set of towns even though their populations and their locations are not radically dissimilar. This is the consequence of the stochastic element in the model which allocates stimuli to one of the towns which, because of the cumulative and circular process of growth, then tends to be at an advantage with respect to its neighbouring towns as its population begins to rise. Also, however,

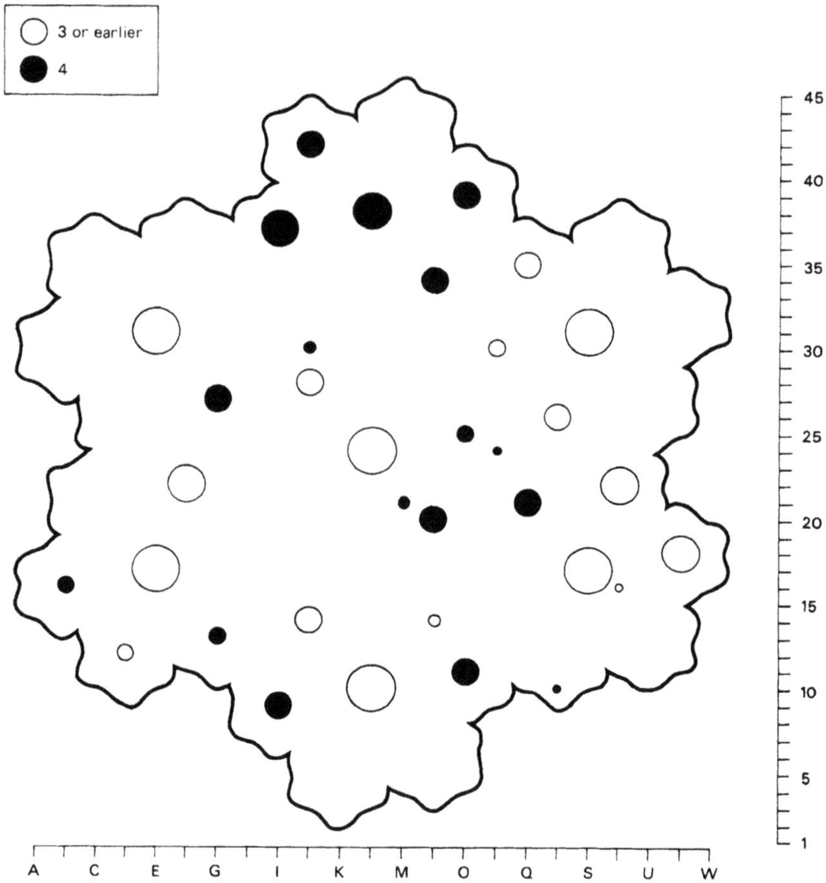

Figure 6.5 The spatial sequence of adoption. Model 2 at generations 3, 4 and 5

one can see that, for all of these places, proximity to the largest town has the effect of increasing the likelihood of earlier adoption and higher growth than in towns of similar size but more remote in the network of the overall set of towns.

Model 2 thus illustrates two points in particular. First, that in dealing with a probabilistic model there is considerable scope for chance happenings to influence the pattern of growth especially in the early time periods of the simulation. Places close to those large centres which happen to adopt early will themselves tend to adopt earlier and grow more rapidly than more distant towns of the same population size, but this is by no means certain. Any particular run of the model will produce a different 'unique' result giving high growth rates to one set of places and lower rates to some other. In all the examples of the application of the simulation model only a single run of each variant of the

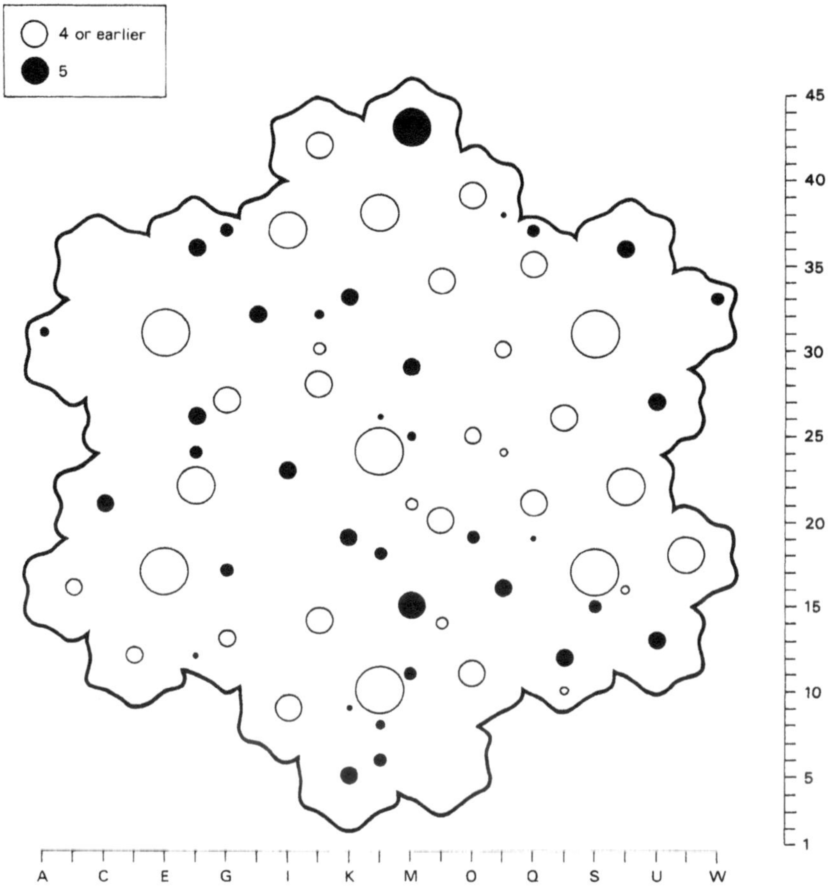

Figure 6.5 (*continued*) Key to sizes of circles is given in fig. 6.2.

model has been made. The interest lies less in the particular pattern which happens to emerge, than in the general tendencies which are suggested and it has been assumed that a single run of the model over a number of time periods can be taken as representative of a 'mean' of a large number of runs.[1] Second, Model 2 illustrates the effect of the joint operation in the diffusion of the composite innovation of the hierarchical and neighbourhood processes and of their consequences on the aggregate pattern of urban growth rates which are produced by growth-causing innovations. The wide scatter of variance for the growth rates of small towns is the product of the neighbourhood effect operating in a set of towns differently located in space. Suppose, for example, that the structure of the diffusion process had been made to depend purely upon population size. The

[1] See, however, the comments in Brown and Moore (1969, pp. 130-1).

effect would have been a radical reduction in the variance of growth rates; any scatter which did exist would have been due solely to the stochastic nature of the simulation. However, in the model, as in the real world,[1] the variance in the distances between small and large places is greater than the variance between large places, and, when a neighbourhood diffusion principle operates, it is this which produces the increasing scatter of adoption dates of small places. To illustrate the different distance variances between large as against small places, consider the distances involved in our idealized landscape of places (fig. 6.1). If the basic unit of length is the distance between one town and any other town, it can be seen that, with respect to the largest place, all of the second-order towns are equally placed at seven units of distance and their variance in this respect is therefore zero. The third-order places, however, lie at different distances from the first-order place; the closest being less than three units of distance away, while the farthest are over nine units distant. The fourth-order places have an even greater variance since their distances from the first-order place vary from only one unit to over ten units. When only the first-order place is sending out stimuli, therefore, the denominator of equation [6.1], which determines the probability of receiving stimuli, will vary by these increasingly wide margins as one moves from second- to fourth-order places. The nature and operation of the neighbourhood effect can therefore be considered to be an agent of some importance in explaining the increasing scatter of urban growth in smaller size-groups of cities. The correspondence between the pattern of real-world growth rates and those produced by the simulation model suggests the plausibility of the argument that the diffusion of innovations within a set of towns is a contributor to the population growth of those towns. This is not to argue that the adoption of innovations can be thought of as the sole 'cause' of urban growth which obviously is the product of a whole host of events of which innovation adoption is but one. It is, however, of more than trivial interest to find that the patterns of urban growth produced by a diffusion model which combines hierarchical and neighbourhood processes is compatible with that of the real world.

VARIANTS OF THE SIMULATION

Model 2 is, of course, only one variant of a whole family of simulation models which would have different operating rules and parameters even though their structure would be identical. There are three principal ways in which the rules could be changed: the calculation of probabilities of receiving stimuli could be changed by altering the exponents of the distance and population terms in

[1] While the vast literature on central-place theory has not established that the uniform spacing and regular hierarchy of that theory are found in the real world, it does suggest the tendency for larger places to be more widely separated than smaller places. See the summary of work in Berry and Pred (1965), and the analysis by E. N. Thomas (1962).

equation [6.1]; the threshold number of stimuli which would be required before adoption occurred could be raised or lowered; and the number of stimuli sent out by towns could be altered. To what extent is it possible to interpret such changes in terms of their meaning in the real world? Altering the distance exponent might be thought of as representing changes in transport in the real world; as movement becomes relatively less expensive or less time-consuming, through technical improvements in transport, so one might imagine that the friction of distance would grow less and the size of the distance exponent would decline.[1] The second change – altering the threshold resistance levels – could be seen in terms of the effects of technical changes upon the optimal scale of production; for any particular innovation, technical improvements might either raise or lower the likelihood of adoption in smaller places either, in the first case, as the risks or the costs of adoption grow less small places can more confidently adopt, or, in the second case, as scale economies increase so will the optimal scale of operation work to the advantage of larger rather than smaller places. The third change – increasing the number of stimuli emitted – could be seen in terms of improvement in general communications. The introduction of electrical means of transmitting messages, the introduction of mass media, improvements in postal services, as well as the improvement of physical transport, all might be expected to increase the total agitation within a national economy and result in an expansion in the overall flow of information and the transmission of ideas (Meier 1962).

It would, however, be difficult to translate into real-world terms any specific change in the operating rules of different varieties of the simulation model since even the three changes noted above are not independent of one another and since the evidence on threshold levels in particular is somewhat ambiguous. On the inter-relatedness of the three examples, improvements in transport, for example, might well have the effect of lowering the distance exponent, but at the same time would doubtless also have the effect of increasing the total flow of information within the system and might too be accompanied by alterations in the threshold levels of adoption. It would, therefore, be difficult to match technical changes during the course of the nineteenth century with specific sets of operating rules so as to develop a version of the model for the early, the middle and the later part of the period 1801-1911, to see whether the changes in the average slope of urban growth rates might be explained by alterations in technology and organization as the period progressed. We might, however, look at two changes in the rules of the simulation to see the direction of their effects and whether these can be made consonant with the empirical changes in urban growth which have been discovered.

Model 3 makes two alterations to the operating rules of Model 2: first by increasing the weighting of the distance term; and second, as a trivial alteration,

[1] See, for example, Olsson 1965, pp. 57-63.

Table 6.4 *Distribution of dates at which places adopt: Model 3*

	Size group							
Period	*1*	*2*	*3*	*4*	*5*	*6*	*7*	*Total*
0	1							1
1						1		1
2	1	1				1		3
3	2			1				3
4		1	2			1		4
5	1	4	4	2	2	1		14
6	1		3	8	4	4	2	22
7			4	6	14	10	4	38
8				6	16	22	10	54
9					4	23	26	53
10				2	8	14	19	43
11						12	21	33
12					2	3	15	20
13						3	16	19
14						2	3	5
15							5	5
After 15						3	22	25
								343

Note: Size groups as in table 6.2.

by making the decline in the population of places which do not receive stimuli a decline of a fixed rather than a proportionate amount. The rules of Model 3 therefore read as follows:

1-3. As in Model 1.

4. The probability of town i being 'hit' by a stimulus is given by the following:

$$k \sum_{j=1}^{n} P_i . d_{ij}^{-2} . S_j \quad (j = 1 \ldots n, j \neq i).$$

5. As in Model 2.

6. If, in the periods after the second cycle of the model, a place fails to receive any stimuli over two consecutive periods, its size will decline by 50 units of population.

The growth rates produced at generation 12 by Model 3 are shown in fig. 6.6. The effect of weighting the distance term by an exponent of 2·0 can be seen by comparison with the results of Model 2. Principally, the difference lies in the

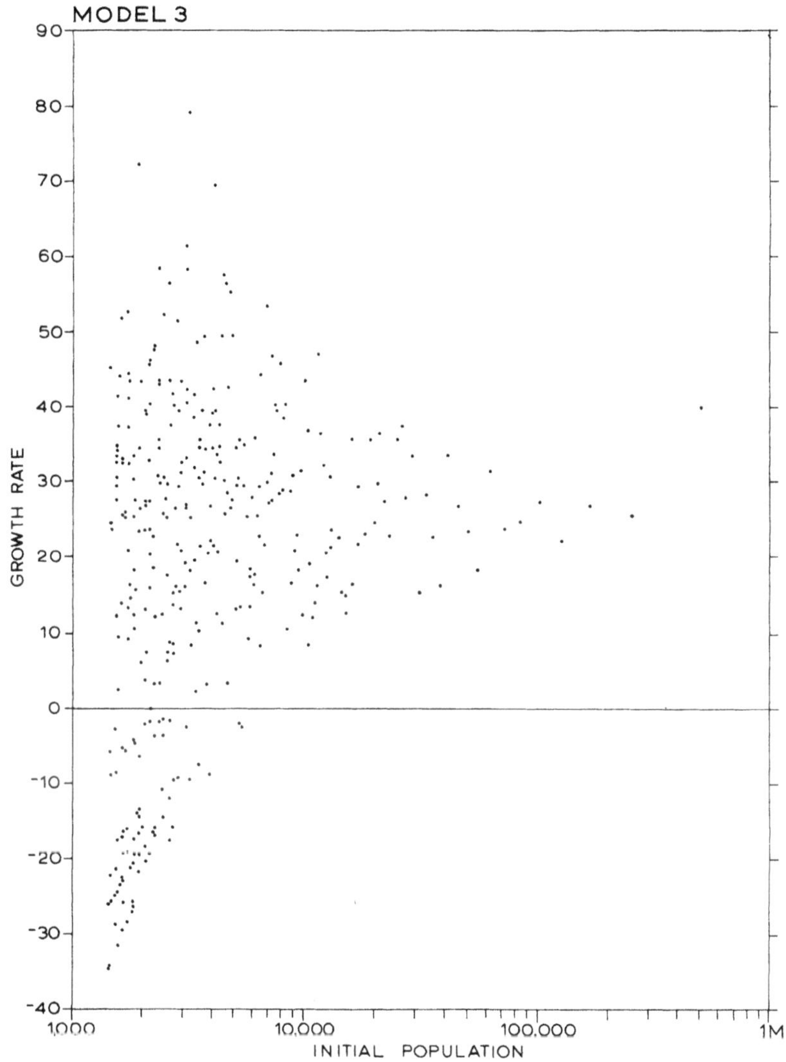

Figure 6.6 Urban growth rates: Model 3 at generation 12

widened scatter of growth rates for each of the size groups in the set of places. Thus, in terms of the date of adoption, table 6.4 shows that there is a much greater spread of dates for the different size groups of towns. This is most noticeable in the group of towns of sizes 2,500-4,999 in which one place adopts as early as period 1 and another as early as period 2. Interestingly, both of these very early adopters are close to the largest place (one is place M.25, the other is

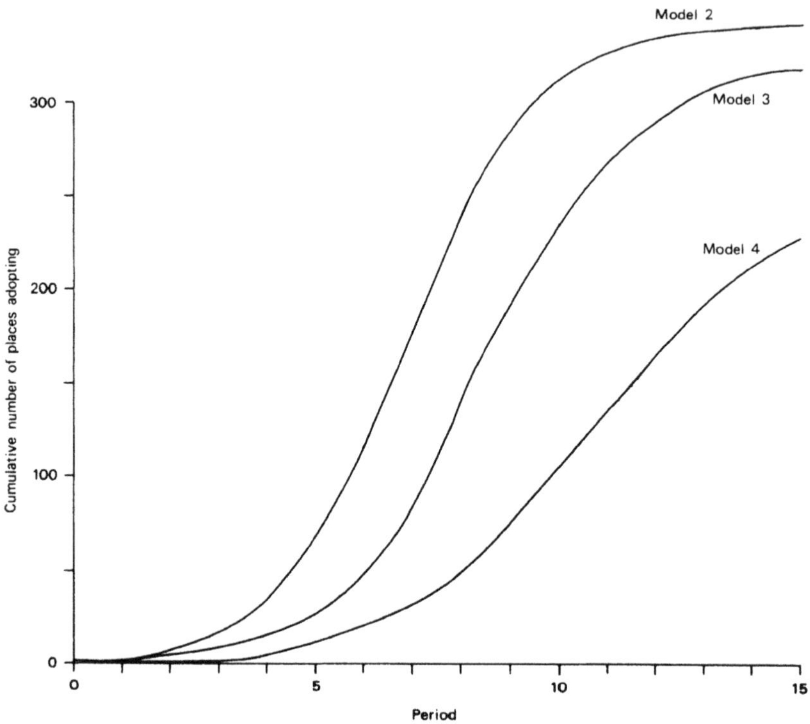

Figure 6.7 Numbers of adopting towns: comparison of Models 2, 3 and 4

place M.23, both of which are small fourth-order towns lying immediately adjacent to the largest town, L.24), thus illustrating the effect of increasing the neighbourhood component of the diffusion process. As was expected, therefore, by giving distance friction a greater weight, certain of the very small places which lie in close proximity to large early adopters benefit more than was the case in Model 2 and the scatter of growth rates in the early periods of the model is consequently somewhat wider. It is apparent too that the greater distance friction also has the effect of slowing up the speed of the whole diffusion process even though the threshold level of adoption is identical for the two models. Figure 6.7 and table 6.5 show the much slower growth curve of adopters in Model 3. The stronger distance constraint has the effect of reducing the probability of stimuli seeping out from the initial adopter town since more of the stimuli are sent to the larger number of small proximate towns each of which takes a longer time than do larger towns to reach the necessary threshold and so adopts later. As the number of adopters is delayed, so the overall growth in the number of stimuli is reduced and a brake is applied to the whole sequence. After

Table 6.5 *Comparison of Models 2, 3 and 4*

	Stimuli sent out			*No. of places adopting (cumulative total in brackets)*		
Period	*Model 2*	*Model 3*	*Model 4*	*Model 2*	*Model 3*	*Model 4*
0				1 (1)	1 (1)	1 (1)
1	27	26	26	1 (2)	1 (2)	0 (1)
2	46	33	26	5 (7)	3 (5)	0 (1)
3	131	71	26	10 (17)	3 (8)	0 (1)
4	251	124	26	17 (34)	4 (12)	4 (5)
5	447	171	99	35 (69)	14 (26)	7 (12)
6	760	347	203	47 (116)	22 (48)	9 (21)
7	1150	559	301	62 (178)	38 (86)	10 (31)
8	1622	889	394	63 (241)	54 (140)	18 (49)
9	2078	1315	572	44 (285)	53 (193)	29 (78)
10	2437	1689	837	28 (313)	43 (236)	22 (100)
11	2700	2029	1022	14 (327)	33 (269)	31 (131)
12	2933	2280	1269	9 (336)	20 (289)	34 (165)
13	3119	2451	1536	3 (339)	19 (308)	28 (193)
14	3301	2638	1743	3 (342)	5 (313)	22 (215)
15	3467	2740	1919	1 (343)	5 (318)	24 (239)
After 15					25 (343)	104 (343)

a number of generations, however, this initial difference is reduced and the two models rapidly converge to produce very similar patterns of growth rates by the tenth period.

A further alteration can be made to one of the operating rules to produce another version of the simulation. Model 4 raises the level of the threshold of resistance amongst the towns from three to five stimuli, and its rules read as follows:

1. As in Model 1.
2. Towns adopt once they have received five stimuli from adopter towns. Having adopted, in subsequent time periods, their populations grow as specified in rule 5 and they emit a certain number of stimuli as in rule 3.
3. As in Model 1.

4-6. As in Model 3.

The growth rates at generation 15 produced by Model 4 are shown in fig. 6.8. The slowing-down in the rate of diffusion is quite dramatic (fig. 6.7 and table 6.5).[1] Almost one-third of the places have not adopted by generation 15, the

[1] It is because of such differences in the rates of diffusion in Models 2, 3 and 4 that the growth rates produced by the three models are shown for different periods in figs. 6.4, 6.6 and 6.8.

Table 6.6 *Distribution of dates at which places adopt: Model 4*

	Size group							
Period	*1*	*2*	*3*	*4*	*5*	*6*	*7*	*Total*
0	1							1
1								0
2								0
3								0
4	3	1						4
5	2	4	1					7
6		1	2	3	2	1		9
7			2	1	4	2	1	10
8			6	4	5	3		18
9			2	8	10	9		29
10				3	7	7	5	22
11				3	7	16	5	31
12				3	7	13	11	34
13					2	14	12	28
14					2	5	15	22
15					1	10	13	24
After 15					3	20	81	104
								343

Note: Size groups as in table 6.2.

period at which cycling of the model was stopped, and, even in the places with populations of from 5,000 to 9,999 three places have not adopted by generation 15. A second difference effected by raising the threshold level, however, is a slight reduction in the scatter of growth rates and adoption dates compared with Model 3 (table 6.6). Again, this is an expected result since it is the consequence of the higher threshold level counteracting the strong distance constraint and thus giving greater weight to population size than in Model 3. With the squared distance term, small places close to a large early adopter are still likely to get some random stimuli because of their proximity and the operation of the neighbourhood effect. However, because there is a large number of such small places, the number of stimuli that any one receives is likely to be relatively few. Thus, with a higher threshold level, these small places are not likely to adopt early and, by later periods, when there is a larger number of stimuli being generated, the more remote small places have had a greater opportunity of catching up the backlog of stimuli so that the proximity advantages of certain of the smaller towns is nullified and they tend to adopt at dates which are not as dissimilar as in Model 3.

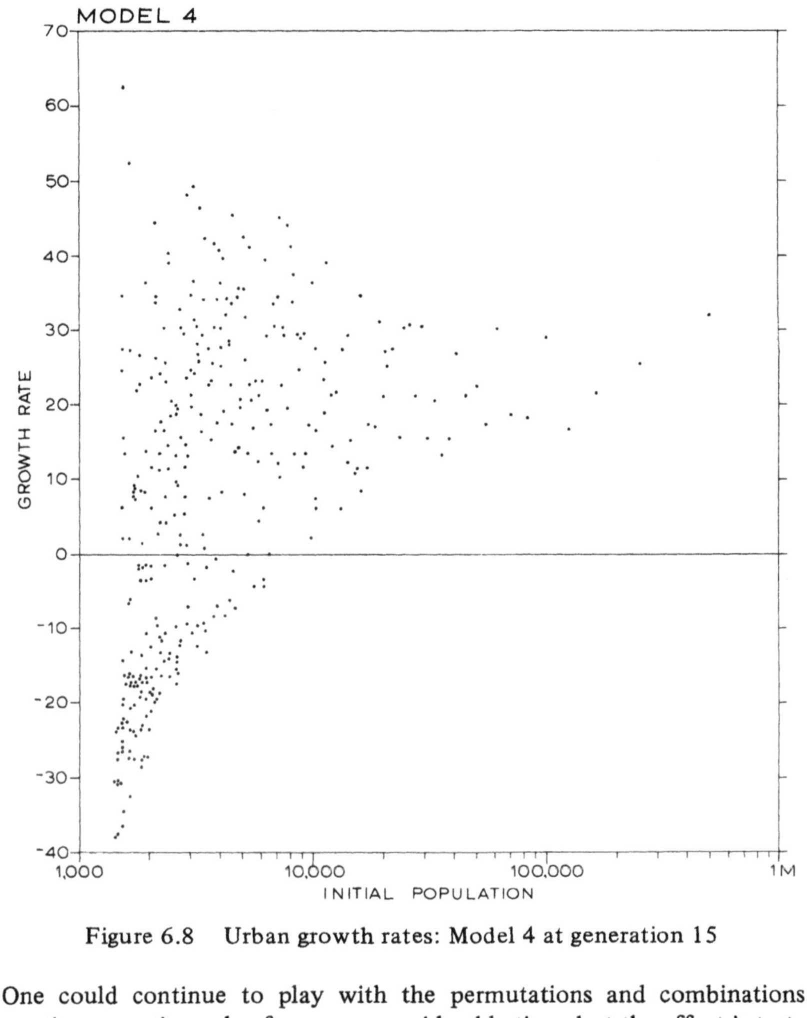

Figure 6.8 Urban growth rates: Model 4 at generation 15

One could continue to play with the permutations and combinations of alternative operating rules for a very considerable time, but the effect is to teach one more about cutting corners in writing computer programmes than about the substantive problem in hand. While a variety of other versions of the basic simulation model were experimented with, sufficient detail has now been introduced for a number of more general conclusions to be drawn.[1]

[1] Further alterations included the use of a variable threshold level as well as different weightings of the population and distance terms in equation [6.1]. The analysis of these alternative models by the highly pragmatic means of actually running simulations is both clumsy and time-consuming, but, given the complexity of dealing with the two dimensions of spatial location in a set of towns, no simple analytic approach suggested itself as an alternative to running the simulations.

CONCLUSIONS

The attempt to simulate the diffusion process of a composite innovation was undertaken in an effort to see to what extent the results on the growth of a set of towns appeared to be consistent with the empirical data discussed in chapter 3. It is apparent that one can reconcile the decreasing variance of growth rates in larger towns with our ideas about innovation diffusion. Because the lines of contact and of stimulus are more distended in the case of small towns, but more especially because the lengths of these links are more *variable* for small than for large towns, there will be a tendency for the effects of the diffusion of innovations to be more variable the smaller the town. Certain small places may well join the ranks of adopters, some indeed may well become adopters at very early dates and thus accrue the benefits hypothesized for the early adopter; but other places will not adopt until very late because, being not only small in size, but also being distant from large centres, they have to await the prior adoption by some neighbouring larger centre before they in turn have a high probability of adoption.[1]

This result is of some wide interest. It provides a rather general explanation for the pattern of variance found in the growth of nineteenth-century towns, even though it is only one of a number of explanations. It also implies that the degree of scatter of urban growth rates may be related to the particular spatial distribution of towns in any set of cities being studied. On the basis of our argument about the importance of the variance of the links of contact in influencing the variance of growth rates, one could hypothesize that a uniform distribution of towns – as in the central-place system – would produce a less wide scatter of growth rates than would a more clustered pattern of towns in which only certain regions were densely settled. Both would still show declining variance at larger city sizes, but the overall variance would be larger for the more clustered set of towns since, in this latter arrangement, all of the large and small places in the densely-settled areas might be expected to adopt early, whereas the towns in the more remote areas of more widely-spaced settlement would be expected to adopt considerably later. It might be anticipated that the variance in the growth rates of all sizes of towns might thus be greater.[2] The spatial pattern of towns in Britain is, of course, somewhat closer to the latter pattern rather than to the formal geometry of central-place theory. The nineteenth-century industrial complexes of the North West, the North East, the Potteries, South

[1] This result is not solely the product of the peculiar and regular geometry of the central-place system from which we started since the populations of the towns were made to vary within each of the hierarchical ranks of central places so that the dispersion of probabilities of receiving stimuli was not solely the product of geometry as it would have been if every town in each order had been allotted equal populations.

[2] It would be a similar process which would provide the rationale for patterns of economic inequality – of core and periphery – to persist within national economies.

Wales, West Yorkshire and the West Midlands were each quite widely separated and each, in turn, distant from London, but within each there was a dense cluster of large and small towns. In such a pattern of places, one would therefore expect the variance of growth rates to be higher than in the simulation model. While all of the simulation trials used an identical basic geometry of places, it would therefore be of some interest to see the extent to which alternative geometries had effect upon the rates of diffusion and upon the consequent urban growth rates.

The second aspect of urban growth which was posed as a problem was the change in the slope of the line relating average growth to town size. From the scatter of growth rates produced by the variants of the simulation model it can be seen that, even by relatively early periods, the different operating rules all rapidly converge on a fairly uniform average growth rate as the number of stimuli grows. In the very early generations however, it is apparent that anything which slows down the overall rate of diffusion works to the advantage of the larger towns and gives them a higher average rate of growth. Both a higher threshold level and a stronger distance constraint therefore have the effect of producing higher average growth rates in the larger towns. The latter of these two influences can certainly be interpreted in terms of improving transport; and, in the context of the nineteenth century, the fact that movement became easier and more rapid would thus be consistent with the progressive change from a set of growth rates positively related to size to one in which size and growth were not related.[1]

However, a much stronger reason can be given for the change in the slope of the curve linking growth and size. It is related to the overall rate at which innovations are introduced into an economy as a whole and here the simulation model is less directly helpful. The model treated diffusions in composite terms. Since the number of stimuli continued to grow with the growth of population and the addition of new adopters, the model made the assumption that the stream of new innovations was an unbroken and ever-swelling flow. Now if it is assumed that major innovations occur in swarms or clusters, as did Schumpeter, and if the effects of each individual innovation becomes progressively less over time, an interesting tendency is apparent. The large towns would show the benefits of their early adoption only so long as new innovations were constantly being introduced. Their 'lead' over small places would continue only so long as they were able to draw on new sets of innovations while the effects of earlier innovations were being diffused to places lower in the size hierarchy. Once the stream of new innovations falters or dries up, however, the larger towns have no further sources of growth on which to draw while, for a time, the smaller places would still be benefiting from their adoption of earlier innovations. This can be

1 This is entirely consistent with Pederson's suggestions in the context of a developing country (Pederson 1970, pp. 224-6).

Table 6.7 *Model 4: distribution of growth rates by generations 10 and 15*

(a) Generation 10 Growth rate (%)	*No. of towns in size category* 7	6	5	4	3	2	1
60– 69							
50– 59							
40– 49	1						
30– 39		2	1				
20– 29		4	4				
10– 19		4	8	6	4	4	3
0– 9	5	12	15	13	9	2	3
–0– –9	6	42	22	6			
–10––19	72	36					
–20––29	50						
–30––39	9						

(b) Generation 15 Growth rate (%)	*No. of towns in size category* 7	6	5	4	3	2	1
60– 69	1						
50– 59	1						
40– 49	2	7	5				
30– 39	7	17	9	4	3	1	1
20– 29	11	19	14	7	6	3	3
10– 19	12	18	13	10	4	2	2
0– 9	16	13	6	4			
–0– –9	13	15	3				
–10––19	42	11					
–20––29	29						
–30––39	9						

Note: Town size categories are as follows
1 80,000+
2 40,000–79,999
3 20,000–39,999
4 10,000–19,999
5 5,000– 9,999
6 2,500– 4,999
7 < 2,500

illustrated by looking at the growth rates produced at successive periods of any one model. Those of Model 4, for example, are shown in table 6.7 for generations 10 and 15. It is apparent that, at the earlier date, the curve relating growth and size slopes upwards since the larger places have higher rates of growth as a consequence of their generally earlier adoption dates.[1] By the later

[1] It is of some interest to note that the one aberrant town with a very early adoption date in the smallest of the size categories is place K.25 – one of the nearest neighbours to the largest town L.24.

date, this difference has almost been eradicated as increasing numbers of the smaller places have come to adopt. If the flow of innovations were then to be stopped, it would be the larger places whose growth would first suffer while the smaller places would still, for a time, benefit from the effects of earlier innovations which were still diffusing through the lower end of the size hierarchy. The maintenance of an upward-sloping growth-size curve thus depends upon there being a continuing stream of innovations from which larger towns can reap, and maintain, the benefits of their generally earlier adoption dates.

The logic of such an oversimplified picture would be that, if one can think of long waves – or Kondratieff waves – within an economy and if the crests of these waves are periods during which new innovations are being introduced, then

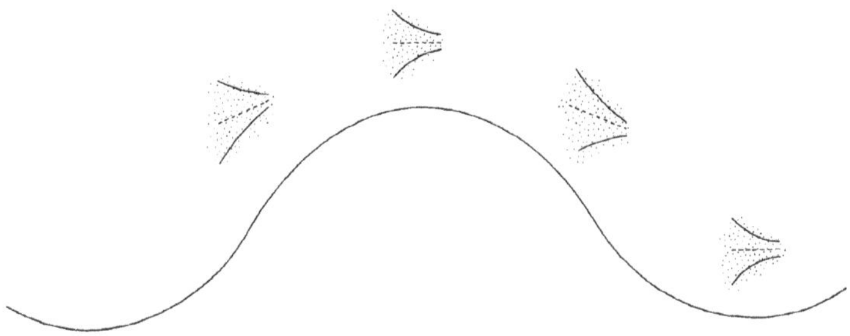

Figure 6.9 The pattern of urban growth and the stream of innovations over time. The solid line suggests an assumed curve of innovations over time in an economy

in these periods of high inventive and innovative activity the large towns would continue to show the benefits of their early adoption of innovations. As the flood of innovations abated so would the advantages of the larger towns begin to decline as the effects of earlier innovations continued in the later-adopting small places while, in the larger towns themselves, no new innovations were available for adoption. The sequence of this idealized relationship might be as portrayed in fig. 6.9 in which the curve of the growth-to-size relationship is shown to alter with the crests and troughs of long waves of innovative activity.

The simulation model of urban growth can therefore suggest reasons both as to why the variance of growth rates grows less in sets of larger cities and why it might be that the relationship of growth and city size may alter over time. Both phenomena can be reconciled with a model suggesting that urban growth is intimately connected with the adoption of innovations which form part of the cumulative and circular process of the growth of an urban economy.

CHAPTER 7

Growth rates and city size: the timing of change

It cannot be said that the country which has produced a Watt, a Stephenson, and an Arkwright is deficient in mechanical inventiveness, but it is impossible to deny that the period of time since 1870 has not been marked by the evolution of distinctly novel electrical inventive ideas proceeding from British minds.

J. A. FLEMING 1901

The diffusion of innovations has been used to explain both of the questions which were posed by the empirical data of nineteenth-century urban growth rates. First, the declining variance of average growth in larger cities has been shown to be compatible with the pattern of growth 'caused' by innovations spreading in a spatial system of towns. Second, the changing relationship between city size and average growth has been suggested as being linked with changes in the flow of innovations; since any cessation in the flow of new innovations in a system of cities could be expected to lead to a relative decline in the average growth rates of larger places, the curve relating size and growth would flatten. The empirical data of the nineteenth century suggest that just such a flattening occurred in the later decades of the period, starting perhaps as early as in the decade 1861-71, but certainly by the decade 1871-81.[1] Since the relationship between size of town and average growth rate changed from being positive to neutral in this period, how reasonable is it to account for the change in terms of a change in the rate of urban innovations?

At first sight, it would appear more profitable to turn to the purely demographic facts of the nineteenth century and seek explanations in them. The rate of population growth certainly fell during the latter part of the nineteenth century, but, for a number of reasons, this does not directly help in any way to explain the change in the relationship between size and growth. First, the fall in the rate of total population growth for England and Wales occurred only in the last two decades of the century and not before 1881 (table 7.1). Secondly, the effect of changes in the overall rate of national population growth would be to change the level of the intercept on the growth-rate axis rather than, by itself, to alter the shape of the curve or the degree of scatter around the average line of

[1] See above, pp. 88–9.

growth. Holding constant the balance of migration, one would therefore expect that, during decades of rapid national population increase, the intercept would be high, and vice versa. Only if the fall in population was differentially concentrated in either large or small cities would the slope of the average growth curve be altered. And the evidence on this question is far from clear. Innes (1938, p. 42) certainly suggests that the fall in the birth rate[1] first occurred in essentially urban occupations, but, on the other hand, it is clear from Welton's (1911, pp. 80-90) calculations of urban population changes, that the fall in fertility in the period 1881-1901 was experienced as early in what he calls 'rural residues' as in the textile and residential towns. Indeed, the fall in the decade

Table 7.1 *Growth of the total population of England and Wales, 1801–1911*

Decade	*Decennial increase* (%)
1801–11	14·00
1811–21	18·06
1821–31	15·80
1831–41	14·27
1841–51	12·65
1851–61	11·90
1861–71	13·21
1871–81	14·36
1881–91	11·65
1891–1901	12·17
1901–1911	10·89

Source: Mitchell (1962, p. 6).

1881-91 was least in large towns, industrial towns and colliery districts (table 7.2). It would therefore seem unproven that birth rates first fell in larger urban areas.

Changes in the flows of migration could, equally, have effect upon the slope of the growth-size curve, since, if migration was predominantly to large towns, then changes in the streams of migration would have differential effect upon towns of different sizes. Such changes did indeed occur during the late nineteenth century, even though, again, they post-date the period at which the change from a positive to a neutral relationship between growth and size occurred. Welton's figures for 1881-1901, for example, show that the textile towns of the north – and, of course, these included some of the very largest of

1 The report on fertility in the census of 1911 suggested that the fall in fertility may be dated from around 1876 and it found that, by 1911, the lowest effective rates of fertility were in the northern textiles towns which combined low fertility with high child mortality (Census 1923, pp. xxxvii and cxxxv).

the towns – suffered considerable losses especially in the decade 1881-91. The towns which predominantly gained most by migration, in contrast, were the residential and military towns. Cairncross (1953, pp. 68-74) likewise suggests that 1880 marked a watershed in the pattern of migration flows to the northern towns; the migrants, who had constituted a not inconsiderable proportion of the increase in the populations of many northern industrial towns, began, after 1880, to move elsewhere, and particularly abroad, instead of to the large industrial towns within Britain. While London escaped such net losses of population, in the northern industrial towns a pattern of net immigration was changed to one of net emigration, both as people from within the towns moved elsewhere and as the streams of migrants from Ireland and the rural areas were diverted abroad in increasing numbers.

Table 7.2 *Differential fertility*

	Birth rates per 10,000 married women aged below 45		
	1881	*1891*	*1901*
Colliery districts	3244	3156	2880
Industrial places	3196	2988	2683
Rural residues	3117	2838	2473
Old towns	2989	2672	2358
Residential places	2932	2553	2188
Large towns	2898	2657	2407
Miscellaneous places	2884	2623	2202
Military towns	2883	2574	2236
Textile towns	2819	2477	2059

Source: Welton (1911, pp. 84 and 87).

An emphasis upon migration flows, however, points as much to consequences as to causes in the changing relationship between growth and size in the last quarter of the century. Not only do the changes in such flows appear to have occurred after the change in this relationship, but, in any case, it has been argued that migration can be regarded as synoptic of the economic and social attractiveness of cities. We therefore need to look beyond the purely demographic changes to suggest causes of the lower relative growth of large towns. What was it about the larger towns of late nineteenth-century Britain which appears to have made them less economically vital, and therefore relatively less able to retain their own population increases or to attract continuing flows of migrants from elsewhere? Amongst a host of aspects, it would not seem unreasonable to focus on two; the economic opportunities offered in large towns and the internal organization of the towns themselves in terms of such things as housing and the array of services and facilities which they offered.[1] Both aspects, of course, bring us back to the question of the adoption of innovations. The first in that flourishing economic opportunities, particularly in the increasingly competitive climate of the time, would depend on the productivity and vitality of trade and industry which would, in turn, depend much upon the types of technique employed and, hence, upon the adoption of new technologies. The second would equally impinge on the adoption of entrepreneurial innovations since the internal functioning of large towns would be much facilitated by organizational innovations which helped to lubricate the complex business of running large towns – supplying water, heat, food, fuel and power as well as making housing available, all would be improved and made possible by the adoption of new techniques in transport, communication and organizational technologies.

The simulation model of urban growth has already suggested that one can 'explain' the change in the slope of growth and size by postulating a fall in the rate of new innovations, since their absence would first have effect upon the growth of larger towns. We are therefore left with the question of why there might have been a fall in the rate of innovations in the final quarter of the nineteenth century: first in innovations in the general sphere of economic activity; and second in innovations affecting the internal organization and operation of large towns.

TECHNICAL INNOVATIONS

On the economic side, there is a large, and still flourishing debate in the literature of economic history, and the ground has been well covered, even though the debate is by no means yet resolved. That contemporaries in the period considered that the country was falling behind, there is little doubt. Levine (1967, p. 11) suggests that by as early as the 1850s doubts were being expressed about the performance of British industry vis-a-vis that of the United States. By the end of the century, there was a welter of commissions and committees set up to investigate such questions and their conclusions almost invariably bemoaned the laggardly British performance. The Royal Commissions on the Depression of Trade and Industry (1884-6), and on Labour (1892-3), the departmental committees organized by the Board of Trade during the First World War, and the studies of individual industries by the Balfour Committee on

1 An interesting question is the extent to which the rising social concern expressed in housing, education and welfare in the final quarter of the century, diverted capital resources into fields which did not repay short-term benefits. Ashworth (1966) asks whether the late Victorian educational and sanitary expenditure brought enough economic benefit to avoid slower economic growth: 'It is at least possible that in the nineteenth century it was more economical to keep the labour force up-to-date by letting early death take care of the old-fashioned, and a high birth-rate, plus the spur of insecurity, provide the new recruitment, than it was to make a social investment for the achievement of longevity and adaptability' (p. 30). He stresses, however, that the social services at the time accounted for a very small fraction of the gross national product.

Industry and Trade after the war, were all joined by the more hysterical voices of contemporary writers such as Shadwell (1909) who added to the mounting criticism of, and unease with, the country's economic performance. Current writing in economic history has explored many of the aspects of industrial retardation which gave rise to such concern. Even though different indices suggest different times for the onset of Britain's decline – indices of industrial productivity, for example, suggest the 1870s, whereas figures of gross national product suggest the 1890s (Richardson 1965a) – there is certainly a general concensus that, with respect to the United States and Germany, British industry showed a marked falling-behind in some, if not all, of the final quarter of the century. The reasons suggested in explanation of this decline are many, but might perhaps be summed under four heads: the early start thesis; the relative factor endowment of the country; rigidities in the organization of its firms; and the market opportunities open to British producers.

The thesis that Britain's early start worked eventually to the country's disadvantage is one of long standing since Veblen (1942) was one of its early advocates. It has, however, been carried further by more recent writing (for example, Frankel 1955 and Svennilson 1954). An important plank of the argument is that the country's early industrial start left it with a stock of equipment which acted as disincentive to its replacement by newer machinery and capital goods and thus reduced the likelihood of new innovations being adopted which would have increased productivity and made goods more competitive. Yet, given that other countries equally had stocks of older equipment, it may be that it was less the capital goods themselves, than the rigid attitudes to methods of production and to social and industrial relations which was the more important residue of the early industrial start. Certainly – as an off-shoot of the early-start thesis – the fact of overcommitment to the basic industries on which the country's first industrial success had been based, had subsequent deleterious consequences. Textiles, railways and consumer goods continued to be favoured at the expense of exports of steel to the developed countries of the world (W. A. Lewis 1957; Richardson 1965b). Continuing overcommitment to these activities had the effect of preempting production facilities which might otherwise have been used for newer types of activity; of perpetuating an industrial structure which became less suited to the new industries so that, even where they were introduced, they were often less economic than elsewhere; and of maintaining the industrially-linked development blocks based on coal, steam and iron which which consequently tended to block the development of electricity and of road transport just as the early chemical industry tended to block the progress of petro-chemicals.[1]

[1] In a not unrelated argument, there are parallels here with Ajo's (1969) thesis that economic systems develop resistance to innovative stimuli and that acceptance of an early innovation tends to reduce the probability of adoption of later stimuli in related fields. He illustrates his argument with the history of Kylälä's invention of an appliance which, fitted to steam locomotives, effected considerable improvements to their heat economy. Adoption of the device, however, was widespread only within France, Spain and Czechoslovakia.

The second argument, that factor endowments had effect upon the relatively poor British performance, relies on the suggestion that, whereas in the United States, for example, labour was relatively scarce by comparison with other factors of production, in Britain there were equally abundant supplies of all factors (except of land), so that there was little incentive to economize on the use of labour by the introduction of labour-saving innovations or to economize on fuel in the operation of, for example, the steel industry or in shipping or railways (see, for example, Habakkuk 1962a). Likewise, looking at the general absence of economies in the use of capital in the tin-plate industry, Aldcroft (1966, p. 129) argues that British industrialists made 'neutral' investment decisions rather than ones with a factor bias, with the consequence that innovations were less rapid than otherwise they might have been.

The third reason adduced by economic historians is the rigidity of the institutional framework in which investment decisions were made. The persistence of small, family-dominated firms checked industrial progress, so it is argued, because of their insular and parochial attitudes and because of the conservatism which tended to characterize their decisions in the generations following the successes of their first founder. Even though, by the turn of the century, a number of large enterprises had come into being in Britain, their scale was in no way comparable with the size of large firms in the United States which had bigger assets, a larger share of the market and covered a broader range of manufactures. Despite the passing of acts which opened the way for the general development of joint stock and limited liability in the period from the 1840s to the early 1860s, the British reaction was a compromise in the form of *private* limited companies many of which still retained the family heads of the older firms. The benefits of limited liability were thus combined with the minimum loss of independence through the retention of family control. This tendency had implications for the effectiveness of many of the amalgamations, since the boards of directors were often comprised of unwieldy bodies formed of all of the previous directors of the amalgamated companies. Payne (1967, p. 529) quotes the case of the Calico Printers' Association with a board of no fewer than eighty-four members: 'These ruinous weaknesses', he comments, 'resulted directly from the original attempts so to organise the combines that each component part retained a high degree of individuality.' Even where relatively large firms did develop in Britain, they differed from their American counterparts in that the reasons for their emergence were often to be found less in the desire to improve performance than to provide buffers against increasing prices and stiffer competition. Such was the case with the amalgamations arising from the 'Brewers' War' late in the century. By 1905, of the largest fifty firms in Britain, eighteen were in brewing or distilling and ten in textile manufacture; while in the United States, only one such firm represented each of these activities and the bulk of the largest fifty firms was in iron and steel, coal and non-ferrous metals – twenty-three firms in all. Of these top fifty firms, in Britain the numbers of 'producer' and 'consumer' firms were twenty-three and twenty-seven respectively; whereas in the United States, the figures were thirty-seven and thirteen (Payne 1967, pp. 527-9).

Small scales of operation had the effect of reducing the likelihood of standardization of production and this is an important component of the fourth reason suggested for British economic backwardness. The market upon which the British manufacturer increasingly relied was in the under-developed countries of the Empire, which offered safety, but little stimulus to encourage technical innovation. The small size of the home market, added to the fact that it had traditionally been supplied with custom-built rather than standardized products, helped to underline the low probability of the introduction of mass-production techniques. In the motor-car industry, for example, the fact that most firms tried to make as many as possible of the component parts for themselves and that they continued to manufacture for a market which expected highly individual products,[1] meant that the notion of inter-changeable parts caught on very slowly in comparison with the United States and France, despite the fact that the industry was better placed than any other to pursue such a technique. 'In Britain the engineers were obsessed by the technical product rather than by the technique of production' (Saul 1962b, p. 43).

Behind many of these arguments lies the notion of an entrepreneurial failure to adapt to changing competition and changing technical expertise in the latter part of the nineteenth century. The debate about entrepreneurial failure in Britain impinges upon much of the discussion rehearsed above and it invokes a number of further reasons for the poor industrial performance, such as the allegedly poorly developed state of technical education in Britain in comparison with the United States and Germany. But the castigation of the British entrepreneur has been somewhat modified since the early writing of Shadwell (1909) and the more recent work of Landes (1969), Aldcroft (1964) and Burn (1940). The various studies of individual industries have softened much of the sting of this criticism and paint the late nineteenth-century entrepreneur as much less of an inefficient sybarite than the earlier, and more general, studies would have led one to believe.[2]

Such writings and catalogues of explanations, while not as systematic as might be wished, do suggest that Britain did indeed suffer by comparison with American industrial productivity at the end of the nineteenth century and they give grounds for suggesting that the rate of adoption of technical innovations was significantly lower. This is not the place to embark upon historical or economic argument, since it is sufficient that we establish that innovations

[1] Such as the habit of manufacturers arranging for customers to fit car bodies of their own design on to a standard chassis.

[2] See, for example, McCloskey and Sandberg (1971) writing, not without irony, in the very journal which gave rise to much of the criticism of entrepreneurial activity. On individual industries, see Saul (1962b, 1967 and 1968) and the contributions in Aldcroft (1968) and McCloskey (1971).

slowed down in late-nineteenth-century Britain and that this could have had differential effect upon the growth rates of the larger towns as against the smaller. Tending to cling to the innovations of the first industrial revolution, British industry did not adopt the wave of newer technologies as rapidly as did certain other countries. The industries which suffered especially in consequence were the steel and chemical industries (for example, Burn 1940; Richardson 1968) and the electrical industries of the 'new' industrial revolution at the end of the century. It was only in the inter-war period that the fuller impact of such newer innovations really took root in Britain so that the impetus of having been the first industrial nation was lost to the United States, Germany and, later, Japan.

ENTREPRENEURIAL INNOVATIONS

While there is some danger in adding endlessly to the list of 'causes' of industrial retardation in the final part of the century, our second emphasis, on entrepreneurial and organizational innovations, does suggest one line of explanation which has not been much followed in the writings of the economic historians. This is the importance of the wider institutional framework in which growth, or lack of growth, occurred. A hostile institutional environment could obviously have marked effects upon the likelihood of innovations being adopted. In particular, stress might be laid upon the importance, in relation to the nature of the innovations at the end of the century, of the size and nature of the control units through which investment, and other decisions affecting innovation adoption, were made. The restrictions which applied to the development of large-scale firms in Britain have already been noted. There is, however, a large number of other fields in which scale of operations was a matter of first importance. The whole question of the servicing of towns at this period depended critically upon the scale of operations. The optimal scale was increasing constantly in terms of a whole variety of servicing activities; from the supply of foodstuffs and retail goods in general, to the supply of gas and other forms of light and power, to the transportation media by which people and information were channelled between and within urban areas. In all of such fields, the nature of late-nineteenth-century innovations laid great stress upon large scale of operation. In the retail trade, this was the period which saw the first developments of departmental stores (Ashworth 1960, pp. 133-7) with the appearance of Marks and Spencer at the turn of the century – with a pattern of diffusion strongly concentrated in the industrial north, the first store opening in Leeds and the first ten including other stores in Warrington, Birkenhead, Castleford, Wigan, Bolton, Manchester and Cardiff – and Woolworths – again with a predominantly northern pattern of diffusion, the first ten stores being in Liverpool, Preston, Manchester, Hull, London, Middlesbrough and Bristol. The development of more extensive supplies of water, gas, electricity, postal services and of internal transport illustrate the importance of scale even more effectively since all required networks encompassing relatively large areas and large numbers of people. The very growth of large towns itself demanded large-scale organization and some increase in the degree of centralization in the control of decision-making.

It might be argued that the functioning of large complexes such as the industrial town might best be achieved in one of two ways: either in terms of a system of *laissez faire* in which few checks were placed on the emergence of a relatively few large-scale private organizations in whose hands many decisions lay; or in terms of a radically-centralized political control by government. At the end of the nineteenth century, Britain was starting to undergo what was to become a major transition towards the latter state, but the lines of battle were only just being drawn in a situation of considerable flux and uncertainty. In the United States, despite the 'trust-busting' exemplified by the Sugar Trust, this was the period of big-business enterprise and one in which the restraints to monopoly had less evident effect. In Germany, by contrast, the state was pre-eminent in a significant number of fields of economic activity. Hartwell (1971, p. 250) suggests that the importance of the differences between English law and continental Roman law may be an important, and certainly an overlooked, element in the timing of the early industrial growth of Britain, but equally it may be argued that it had effect in the later period too since Roman law lent itself much more readily to centralized bureaucratic control and to the large-scale state involvement which characterized many of the continental countries. At the end of the century, when large scale was becoming of growing importance, this might have worked to the benefit of such areas.

Put crudely, therefore, one might argue that, unlike the development of large-scale firms in the United States or the large-scale state involvement in certain European countries, the late nineteenth century in Britain was a period in which no well-crystallized pattern of large-scale control had developed, but one in which the debate about the most optimal units of control was being vigorously, if somewhat damagingly, debated. It was a period of transition in which the old order of mistrust of central authority and a belief in the virtues of free trade, was giving way to the development of a centralized and welfare state. The debate over centralization had, of course, continued throughout the nineteenth century,[1] but it was coming to a head in the late nineteenth century both in the reform of political boundaries and, of more significance, in the advocacy of state involvement in a variety of economic and social matters. There was, too, a notable contribution to this debate in the growing voice of the municipalities which, while formed in the 1830s, had begun to flex both their

[1] Earlier reaction to the Poor Law legislation of the 1830s and '40s, to the Factory Acts and to the railway legislation of the 1840s is discussed by Lubenow (1971) who charts the hostility which was expressed towards what were seen by many as instances of an undesirable growth in governmental control in the early Victorian period.

political and economic muscles in the period after 1870. The debate was thus a twin affair; authority versus industry and central versus local authority. The uncertainty of control, and the competition arising from there being a multitude of potential mechanisms of control, undermined much of the social legislation of the late nineteenth century, but it also had profound effects upon the fate of many entrepreneurial innovations at a time when many of the new inventions and innovations required large-scale control and application. The innovations of the electrical age, with their emphasis on size and on wide-ranging networks, coincided in Britain with the emerging, but not yet fully realized, transfer of power and control to central authorities. The clash between the scale of operations demanded by the new wave of innovations and the uncertain scale of control of decision-making may well have been an element of considerable significance in accounting for the lags in the adoption of the new innovations of the 'second industrial revolution' in late-nineteenth-century Britain.

The examples which could be cited by way of illustration are numerous. The lag in the adoption of electricity itself is one. So far as the use of electricity in power and lighting were concerned, the municipal control of gasworks meant that there was considerable opposition from municipal authorities to the adoption of what was rightly seen as a powerful competitor threatening to undermine a source of municipal revenue. And to this was added the restrictive legislation of the Act of 1882 which must itself be seen both as a sign of the temerity of the government as well as its acquiescence to fragmentary municipal control rather than to potentially monopolistic and unitary private control (see, for example, Garke 1907). The effect was that British electrical lighting lagged seriously behind developments in the United States (Hughes 1962) with consequences upon the embryonic electrical industry as a whole (Du Boff 1967) and upon the myriad of industries to which it was potentially linked. The adoption of electricity in Britain would, indeed, have been considerably slower than it was had not American and German firms stepped in at the turn of the century to provide much of the necessary machinery and equipment; the American and German firms of Westinghouse, General Electric, A.E.G., and Siemens contributed greatly to the early installations as did the American firms which were actually established within Britain in the period after 1890.[1]

A second illustration is provided by the lag in internal transport within British cities. The development of tramways was again seriously hampered by the competition between private and municipal desire for control, and supporters of the municipal cause were partly motivated by the uncertain attitudes of the time towards monopoly and large-scale control (Leech 1897-8). The notion that competition was the great economizer and that this could be translated in the

[1] An interesting sidelight on the question of scale is provided by the experience of one of these immigrant firms. Ferranti's huge power station which was established at Deptford failed commercially partly through the Board of Trade's policy of establishing very small supply areas (Byatt 1968).

case of private undertakings into the encouragement of developments by more than one company within a single town, made no more sense in the context of internal transportation than it had, much earlier, in the case of the duplication of railway lines. Meyer's (1906) somewhat frenetic denouncements of the evils of municipal control of tramways was in essence an argument for lack of restrictions and for the merits of single- as against multiple-working of such enterprises, and they contain much sound common sense.

These are but some of the many possible illustrations of the retarding effects of the discordance between the necessary scale of operations of certain of the new innovations of the late nineteenth century and the scale at which control and decisions could readily be effected at the time. We might look further at the question by drawing on the history of yet another of the examples – the development of the telephone, whose diffusion was studied in an earlier chapter. This is but one of a multitude of possible examples which illustrate the effect of an inappropriate scale of operations upon the spread of innovations.

SCALE OF OPERATION: THE CASE OF THE TELEPHONE

That the spread of the telephone in Britain lagged behind that in America and other European countries, there is little doubt. Webb (1910, p. 21) calculated that, at the start of 1910, Britain had just over 600,000 telephones and that, had it been as well provided as the United States, it would have had over 2,500,000.[1] *The Electrician* used the theme of the unequal provision of the two countries as a recurring topic of debate, commenting on the disparate sets of statistics with asides of varying acidity: 'Go on America. Tell us the work you are doing and we will follow *afar off*' (*Electrician* 5, 1880, 266). It was claimed that, by the beginning of 1880, America had exchanges in operation in no fewer than eighty-five towns (*Electrician* 4, 1880, 73) and that, by the end of 1881, only one town of those with populations greater than 15,000 and only nine of those with populations greater than 10,000 did not have exchanges (*Electrician* 7, 1881, 178). By the early years of this century, it was claimed that the United States added an annual total of telephones twice the total number in existence in Britain (*National Telephone Journal* 1, 1906, 168).

Two aspects of the development of the telephone in Britain illustrate the types of barrier put in the way of its expansion – barriers which were closely connected with the twin aspects of scale of operation and control of decision-making: first was the role of the state; and second, that of the municipal authorities. The state, given its expensive involvement with the telegraph service which it had taken over from private companies in 1870, had a vested interest in hindering the development of a potential rival in the field of general communications, even though the full extent of that rivalry was not at first

[1] Further comparative statistics are given for European cities, for example, by C. E. Webber (1882) and, for America, by Baldwin (1925, pp. 616-17).

realized.[1] State involvement began early with the Post Office challenge that the telephone was but another form of telegraph and so infringed the state monopoly in telegraphs. It was the settlement of the ensuing court decision in favour of the Post Office which gave rise to the compromise by which the companies were made licensees of the Post Office on payment of a 10 per cent royalty on their receipts. This compromise was characteristic of the continuing uncertainty in the government's attitude to the telephone companies. It led to its rather damaging dilatoriness in making an eventual decision to take over the private companies in the early years of this century, more than thirty years after the first introduction of the telephone. The effect, in the meantime, had been that, not only were many companies refused licences to operate, but that the activities of the companies were constantly shackled by the uncertainty which surrounded the long-term future of their investments. Talk of a government take-over had existed from the very early years: even in 1879, consideration of government control of the telephone was being urged (*Electrician* 3, 1879, 258); rumours that the government was seriously considering such a step were being reported in the following year (*Electrician*, 5, 1880, 229); and, again, there were reports in the press that the Postmaster-General had claimed that the government was about to buy-out the companies in 1888 (*Electrician* 20, 1888, 310), reports which were subsequently denied, but which nevertheless added to the uncertainty of the position of the companies. The original telephone licences had been issued so as to expire at the end of thirty-one years after the start of the early licensing system in 1881, but the certainty of take-over and the terms upon which it were to be effected, were not formally agreed until the early part of this century – 1901 in the case of London and 1905 elsewhere. The agreement was for a state assumption of control at the start of 1912 and, since the system was to be bought at replacement cost without allowance being made for goodwill or custom, for the intervening period the National Telephone Company had little incentive to develop its system much farther than that which already existed.

The government's hesitation over assuming control of the telephone was damaging to the development of the service and it can be accounted for not only in terms of its existing vested interests in the telegraph, but also in terms of the prevailing belief in the health of competition and the danger of monopoly. Large companies were suspect, even though the nature of the telephone demanded highly centralized control. The amalgamation of the various private companies into the National Telephone Company by the middle of the 1890s, was greeted with considerable mistrust and, from the early 1880s, the Post Office had both opened exchanges of its own as well as having granted licences to small rival private companies. While the extent and success of the Post Office exchanges were very limited, except perhaps in the North East, the fact that it was seen as

[1] See above, pp. 175–6.

desirable to establish rival systems, suggests the belief in the virtues of competition in spite of its evident inappropriateness in the case of a system which was more efficient the greater was its network character and the greater was the extent of that network. By offering alternative telephone systems, the Post Office, or the municipalities, or competing private companies merely duplicated facilities and forced customers to subscribe to more than one network. It was such illogicality which lead inexorably to the gradual emergence of but a single company, but the moral – whose recognition was much delayed by the belief in competition – should have been that, in whatever hands it were to lie – whether state or private company – monopoly or large-scale control was the order of the day where such innovations based on electricity were concerned.

The American experience illustrates a different, and more appropriate, scale of operations. The National Bell Telephone Company had emerged as an amalgamation of two earlier companies by 1879; by 1885, the American Telephone and Telegraph Company had been formed to organize long-distance communications; and, by 1899, this company had merged with Bell to provide a highly centralized administrative system for its associated manufacturing and service companies as well as providing facilities for research.[1] Baldwin concludes that, 'The telephone business in America thus grew to be a huge undertaking long before it became a concern of any magnitude in this country, and it therefore merited an organization and expenditure in research and development which it has been denied here' (Baldwin 1925, p. 615).

The second aspect of the development of the telephone which illustrates the theme of the scale of control units is the involvement of local authorities. Responsible, as they were, for single municipalities their influence was on the whole somewhat parochial and certainly at a scale too small to match the requirements of an evolving network of development. Meyer (1907, p. 266) drew attention to this small-scale parochialism of municipal involvement in urban electrical undertakings in his suggestion that local authorities 'were unable to cooperate for the purpose of serving conjointly a number of districts which the ties of manufacture had united into a commercial and manufacturing unit'. Again, while not painting too crude a dichotomy between the American and British situations, the situation between the two countries was made somewhat different by the less strong position of municipal bodies in the more raw political climate of the United States. In Britain, the provision of way-leaves was one of the recurring and most serious problems hindering the expansion of the telephone and municipal opposition to granting way-leave rights was widespread and consistent. Their refusals to grant extensions to overhead wires were often fully justified since, as well as being unsightly and dangerous, the overhead wires

[1] In a wider context, Edison's establishment at Menloe Park in 1871 of a comprehensive research laboratory can be seen as the first development of the concept of centrally-organized technological research.

were increasingly inefficient in the later years of the telephone. But the municipal refusal to grant permission for underground cables can be seen almost as an exercise in flexing their new-found municipal muscle. In the case of both telephones and tramways, the municipalities developed what virtually became a philosophy of the municipal sanctity of the streets in their control – a philosophy which was extended to the air above the streets as well as to the breaking-open of the actual surface. Their attitude was also motivated by the more mundane consideration of the potential income source which telephone development offered. It is not without significance that at the head of many of the deputations demanding lower telephone rates were representatives of the new London County Council, of the Association of Municipal Corporations and of the Scottish municipalities. The Association of British Chambers of Commerce, founded in 1870, also played a part in such movements. It was pressure from such bodies that led to the eventual agreement, by government legislation in 1899, to grant licences to any local authority which applied to operate a municipal telephone exchange. The six municipal exchanges which were actually established as a result of this pressure acted as irritants rather than as serious competitors to the National Telephone Company, since all but the Hull exchange rapidly disappeared, most having been bought by the Company at a loss to the municipalities and never having worked at a profit, being regarded 'by experienced engineers as obsolete even at the time they were installed' (Baldwin 1925, p. 381).

Even though their realization proved so disastrous a failure and even though the government itself regarded the operation with some misgivings, the fact of the pressure for municipal exchanges does illustrate the effects of the contemporary debate on what was the optimal unit of political and economic control in this transition period. Fabian tracts argued for municipal control of a wide variety of undertakings; the heirs of Bentham and of Chadwick argued for greater central control; and the popular spirit lauded the virtues of competition and deplored monopoly in ways at once inconsistent and pragmatic. And in the midst of such uncertainty and of the fluctuating opinions which so characterized the early as well as the later Victorian age, those innovations which increasingly required large-scale control, regardless of whether it was in the hands of private or public agency, languished in the confusion. The nature of so large a proportion of the innovations of the 'second industrial revolution', based as they were on electricity and on networks, demanded a scale of unified control which the current ferment of debate at the end of the century did not provide, even though it heralded the eventual evolution of more centralized and large-scale control.

Most discussion of urban growth has concentrated upon the role of demographic and economic factors, but it is perhaps equally important to study the effect of this factor of control and decision-making, which might be exemplified both in the scale and the intensity of political institutions and in attitudes towards them. In this study, it has been possible to show the importance of the diffusion of innovations upon the patterns of urban growth, just as other studies have shown the impact of innovations upon general economic growth. But in looking at this process, the role of constitutional theory and political power have been highlighted. The tantalizing glimpses of the influence of the struggle between central and local government and between the state and the individual upon the spread of innovations suggest new routes of explanation for the geographer. The scale and nature of power and of the locus of decisions – whether in terms of political power, the activities of manufacturing corporations, or of other sources – have been sadly neglected by geographers, despite their evident impact upon the more tangible aspects of the economic landscape.

In what ways do differences in the real or the perceived structure of power and decision-making vary from place to place and over time, and in what ways do such differences have effect upon the economic vitality and upon the innovativeness of the areas and times concerned? Real differences certainly exist. In terms of the structure of local government at the present, one need only look at the varying stances adopted towards the concept of comprehensive education, at approaches to the management of public housing or at the varying proportions of local finance which are allocated to particular social services, to appreciate the breadth of variation in the interpretation and the exercise of the local government mandate. In what ways do such differences – and in the perception of such differences – affect the buoyancy of local economies? The question takes on added interest at the present, in the context of the current major reform of local government. This reform offers great scope for the analysis of such variation, not only in relation to the efficiency of new local government boundaries and of the appropriateness of the scale of the new units in relation to the various responsibilities which fall within the province of local government, but also on the broader front in relation to the effectiveness of the proposed structure of local government as a mediator between the state and the individual. The major reshuffling of local government in the 1880s, from which the strength of local government derived, can be seen as a compromise solution to the competing claims of *laissez faire* and centralization. Britain today is still left with the legacy of that compromise. To what extent does it better serve the scale requirements of innovations in such varied fields as education, housing, transport, industry and retailing than do other more highly centralized systems on the one hand or more individualistic systems on the other?

APPENDIX A

The simulation model: urban growth and innovations

In massaging a body of data which covers almost 1,000 towns over a span of eleven decades, a great deal of computer calculation was obviously involved. Throughout the text, the details of statistical techniques and of quantitative methods have been presented in as simple a fashion as possible. For those interested in the details of the simulation model, however, the computer programme which was used is given below. The language used in writing the programme is FORTRAN IV and the programme calls for only one major subroutine – the derivation of uniformly-distributed random numbers.

The data requirements of the programme, for each place, are as follows: a reference number (NAME), an initial population size (XPOP), and locational co-ordinates (XLOC and YLOC). The version given below refers to Model 4 whose operating rules are specified in the text (p. 207). The programme, as written, cycles for fifteen generations and will print out the following: for generations 1 to 4, a list of the number of 'stimuli' received by each town; for generations 5, 10, 11, 12 and 15 the initial population, new population, growth rate, date of adoption, number of stimuli received and probability of being 'hit' for each of the 455 places. Total running time, using an IBM 360/44, is approximately eleven minutes. This can be considerably reduced by calling on a pre-calculated matrix of distances between each of the 455 places used in the simulation, but, for simplicity, the programme shown here includes the calculation of distances as an integral part of the programme (lines 34 to 36).

```
1          SUBROUTINE RANDU (IX, IY, YFL)
2          IY=IY*65539
3          IF(IY)5,6,6
4       5  IY=IY+2147483647+1
5       6  YFL=IY
6          YFL=YFL*0.4656613E-9
7          RETURN
8        END

1            DIMENSION XPOP(455),POP(455), XLOC(455),
           5  YLOC(455),NAME(455),STIM(455),HITS(455),
           5  ITOHIT(455),EXHITS(455),IDATE(455),IPROB(455),
           5  PROB(455),IBLIM(456),GRATE(455)
```

```
  111  FORMAT('',I5,I8,2F12.0,F16.4,3I10)
  222  FORMAT('0','STIMULI SENT OUT IN',I3,
     5  'TH GENERATION=',I6)
  333  FORMAT('0',10X,'CITY',3X,'INITIAL POP',3X,'NEW POP',
     5   4X,'GROWTH RATE',4X,'DATE HIT',4X,'NO. HITS',
     5   4X,'PROBS',I6,'TH PERIOD')
  666  FORMAT(I3,2F3.0.F7.0)
  777  FORMAT('',I5,I8,2I10)
  999  FORMAT('0','DIFFUSION OF GROWTH STIMULI-
     5   MODEL 4')
       DO 3 I=1,455
       READ(5,666) NAME(I), XLOC(I), YLOC(I),XPOP(I)

C  INITIALIZE ELEMENTS BEFORE STARTING MAJOR
C     GENERATION CYCLES
       POP(I)=XPOP(I)
       ITOHIT(I)=0
       HITS(I)=0.0
       XLOC(I)=XLOC(I)*8.6605
       YLOC(I)=YLOC(I)*5.0
       IDATE(I)=0
     3 CONTINUE
       WRITE(6,999)

C START PLACE 1 WITH VALUES APPROPRIATE TO
C    ADOPTION
       ITOHIT(1)=5
       STIM (1)=26.5
       ISTIMT=26

C  ALLOT RANDOMLY-CHOSEN VALUES TO ARGUMENTS
C     IN RANDOM SUB-ROUTINE
       IYA=24109
       IXA=567
       YFLA=0.123

C START MAJOR CYCLES OF GENERATIONS
       DO 100 JI=1,15
       PRTOT=0.0
       DO 10 I=1,455
       EXHITS(I)=HITS(I)*0.5
       HITS(I)=0.0
       IPROB(I)=0
       PROB(I)=0.0

C  CALCULATE PERIOD AT WHICH EACH PLACE ADOPTS
       IF(ITOHIT(I).LT.5) IDATE(I)=IDATE(I)+1

C   CALCULATE DISTANCES BETWEEN ADOPTERS AND ALL
C      OTHER PLACES
       DO 20 K=1,455
```

```
          DIFFX=XLOC(I)-XLOC(K)
          DIFFY=YLOC(I)-YLOC(K)
          DIST=(DIFFX*DIFFX)+(DIFFY*DIFFY)
          PROB(I)=PROB(I)+(POP(I)*STIM(K)/DIST)
       20 CONTINUE
C   CALCULATE TOTAL 'PROBABILITIES' OF ALL PLACES
          PRTOT=PRTOT+PROB(I)
       10 CONTINUE
          IULIM=0
C   CONVERT PROBABILITIES TO INTEGERS SUMMING TO
C     1,000,000
          DO 30 I=1,455
          IPROB(I)=IFIX((PROB(I)/PRTOT)*1000.0*1000.0)
C   EXPRESS INTEGER PROBABILITIES AS FIELD OF
C     CONSECUTIVE NUMBERS
          IF(IPROB(I)-1) 55,66,66
       55 IBLIM(I)=0
          GO TO 30
       66 IBLIM(I)=IULIM+1
          IULIM=IULIM+IPROB(I)
       30 CONTINUE
          IBLIM(456)=IULIM+1
C   GENERATE ISTIMT RANDOM NUMBERS BETWEEN 0 AND
C     1,000,000
          DO 50 L=1,ISTIMT
          IXA=IYA
          CALL RANDU(IXA,IYA,YFLA)
          IRAND=IFIX(YFLA*1000.0*1000.0)
C   IF RANDOM NUMBER IS 0, GENERATE AN ALTERNATIVE
C     NUMBER
          IF(IRAND-1) 88,99,99
       88 ISTIMT=ISTIMT+1
          GO TO 50
C   ALLOCATE RANDOM NUMBERS TO THE FIELD OF
C     CONSECUTIVE PROBABILITIES
       99 DO 50 K=1,455
          IF(IBLIM(K).EQ.0) GO TO 50
          IF(IRAND.GE.IBLIM(K).AND.IRAND.LT.IBLIM(K+1))
          5 GO TO 5
          GO TO 50
        5 HITS(K)=HITS(K)+1
          ITOHIT(K)=ITOHIT(K)+1
       50 CONTINUE
          WRITE(6,222) JI, ISTIMT
```

```
      DO 80 M=1,455
C POPULATION DECLINES IF, AFTER 2ND GENERATION,
C    PLACE IS NOT HIT DURING TWO CONSECUTIVE
C    PERIODS
      IF(JI-2) 101,101,201
  201 IF(HITS(M).LT.1.0. AND.EXHITS(M).LT.0.5)
     5   POP(M)=POP(M)-50.0

C CALCULATE NEW POPULATIONS AND STIMULI FOR ALL
C    OTHER PLACES
  101 IF(ITOHIT(M).LT.5) GO TO 80
      STIM(M)=SQRT(SQRT(POP(M)))
      POP(M)=POP(M)+((HITS(M)+1.0+EXHITS(M))*100.0)
      ISTIMT=ISTIMT+IFIX(STIM(M))
   80 CONTINUE
      IF(JI.LT.5) GO TO 12
      IF(JI.EQ.5.OR.JI.EQ.10.OR.JI.EQ.11.OR.JI.EQ.12.OR.
     5   JI.EQ.15) GO TO 11
      GO TO 100
   12 WRITE(6,777) (I,NAME(I),IPROB(I),ITOHIT(I), I=1,455
      GO TO 100

C CALCULATE GROWTH RATES FOR SELECTED GENERATIONS
   11 WATE=2000.0/FLOAT(JI)
      DO 22 I=1,455
      GRATE(I)=((POP(I)-XPOP(I))/(POP(I)+XPOP(I)))*WATE
   22 CONTINUE
      WRITE(6,333) JI
      WRITE(6,111) (I,NAME(I),XPOP(I),POP(I),GRATE(I),
     5   IDATE(I),ITOHIT(I),IPROB(I), I=1,455
  100 CONTINUE
      STOP
      END
```

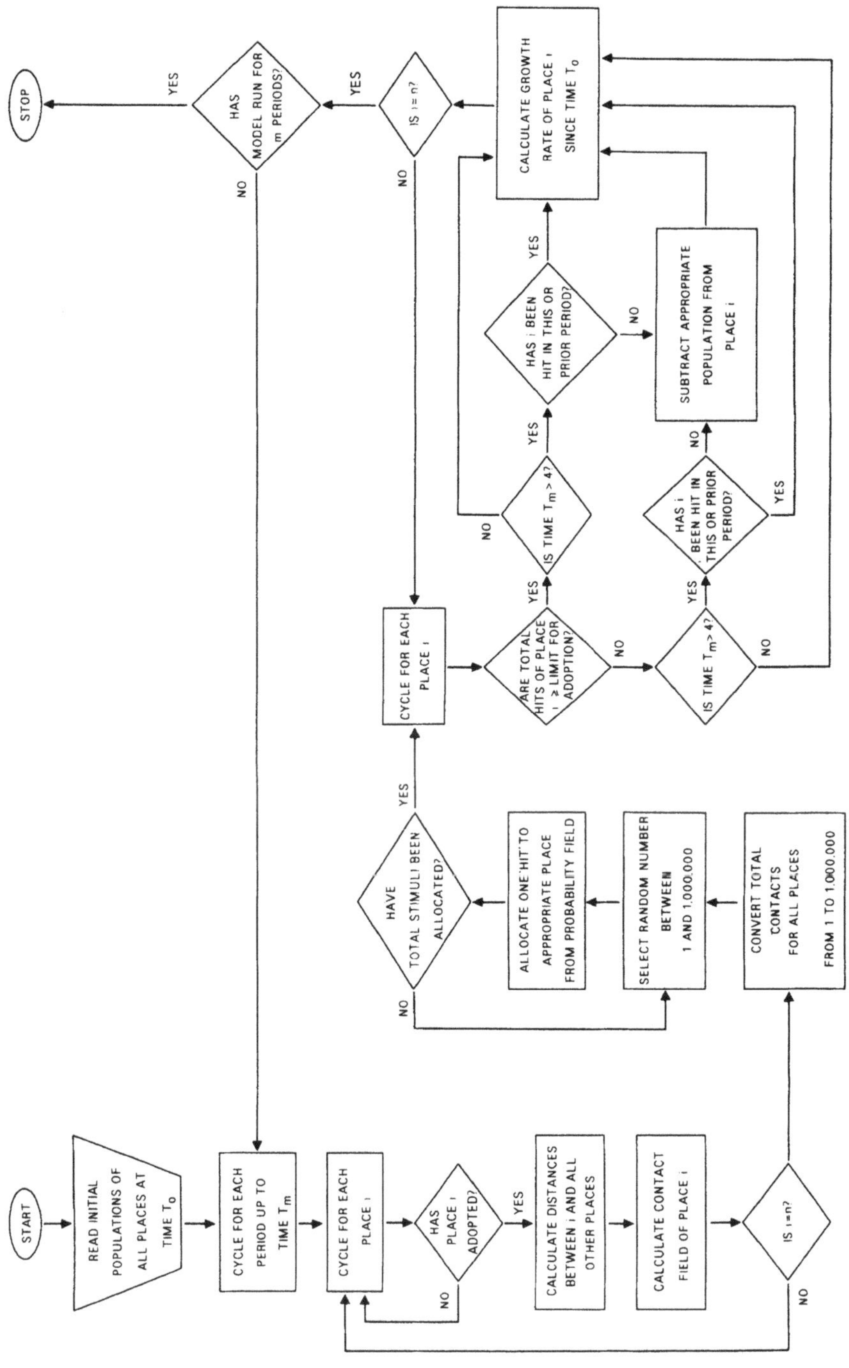

Appendix A Flowchart of computer programme for the urban simulation model

APPENDIX B

List of towns

(Locations of towns are shown on the reference map, pp. 242–3)

Bedfordshire (BED)

1 Bedford
2 Biggleswade
3 Dunstable
4 Leighton Buzzard
5 Luton
6 Sandy

Berkshire (BER)

1 Abingdon
2 Maidenhead
3 Newbury
4 Reading
5 Wallingford
6 Wantage
7 Windsor & Eton
8 Wokingham

Buckinghamshire (BUK)

1 Aylesbury
2 Bletchley
3 Buckingham
4 Chesham
5 High Wycombe
6 Marlow
7 Newport Pagnell
8 Olney
9 Slough
10 Wolverton

Cambridgeshire (CAM)

1 Cambridge
2 Chatteris
3 Ely
4 March
5 Whittlesey
6 Wisbech

Cheshire (CHE)

1 Alderley Edge
2 Alsager
3 Altrincham
4 Bollington
5 Bredbury
6 Cheadle
7 Chester
8 Congleton
9 Crewe
10 Ellesmere Port
11 Frodsham
12 Hazel Grove
13 Heswall
14 Hollingworth
15 Hoylake & West Kirby
16 Hyde
17 Knutsford
18 Lymm
19 Macclesfield
20 Marple
21 Middlewich
22 Nantwich
23 Neston & Parkgate
24 Northenden
25 Northwich
26 Runcorn
27 Sale
28 Sandbach
29 Staleybridge
30 Stockport
31 Wilmslow
32 Winsford

Cornwall (COR)

1 Bodmin
2 Camborne
3 Falmouth
4 Hayle
5 Helston
6 Launceston
7 Liskeard
8 Looe
9 Newquay
10 Penzance
11 Penryn
12 Redruth
13 Saltash
14 St Austell
15 St Blazey
16 St Ives
17 Stratton & Bude
18 Truro

Cumberland (CUM)

1 Aspatria
2 Carlisle
3 Cleator Moor
4 Cockermouth
5 Egremont
6 Frizington
7 Harrington
8 Keswick
9 Maryport
10 Millom
11 Penrith
12 Whitehaven
13 Wigton
14 Workington

Derbyshire (DER)

1 Alfreton
2 Ashbourne
3 Bakewell
4 Beighton
5 Belper
6 Bolsover
7 Brimington
8 Buxton
9 Chesterfield
10 Church Gresley
11 Clay Cross
12 Clowne
13 Derby
14 Dronfield
15 Elmton

16 Eckington, Mosborough & Renishaw
17 Glossop
18 Hasland
19 Heage
20 Heanor
21 Ilkeston
22 Killamarsh
23 Long Eaton
24 Matlock
25 Melbourne
26 New Hall
27 New Mills
28 Pilsley
29 Pinxton
30 Ripley
31 Shirebrook
32 South Normanton
33 Spondon
34 Staveley
35 Tibshelf
36 Whitwell
37 Whittington
38 Wirksworth
39 Woodville

Devon (DEV)

1 Barnstaple
2 Bideford
3 Brixham
4 Buckfastleigh
5 Crediton
6 Dartmouth
7 Dawlish
8 Exeter
9 Exmouth
10 Great Torrington
11 Honiton
12 Ilfracombe
13 Kingsbridge
14 Newton Abbot
15 Okehampton
16 Paignton
17 Plymouth
18 Sidmouth
19 Tavistock
20 Teignmouth
21 Tiverton
22 Topsham
23 Torquay
24 Totnes

Dorset (DOR)

1 Blandford Forum
2 Bridport
3 Dorchester
4 Lyme Regis
5 Poole
6 Portland
7 Shaftesbury
8 Sherborne
9 Swanage
10 Weymouth
11 Wimborne Minster

Durham (DUR)

1 Annfield Plain
2 Barnard Castle
3 Birtley
4 Bishop Auckland
5 Blaydon
6 Boldon Colliery
7 Brandon & Byshottles
8 Carrville
9 Chester-le-Street
10 Chilton
11 Consett
12 Cornforth
13 Coundon
14 Coundon Grange
15 Crook
16 Darlington
17 Durham
18 Ferryhill
19 Hartlepools
20 Hebburn
21 Hetton-le-Hole
22 Houghton-le-Spring
23 Hylton
24 Jarrow
25 Langley Park
26 Leadgate
27 Murton
28 Newbottle
29 Pelton
30 Penshaw
31 Ryhope
32 Seaham
33 Seaham Harbour
34 Sherburn
35 Shildon
36 Shotton
37 South Hylton
38 South Shields
39 Spennymoor
40 Stanley
41 Stockton-on-Tees
42 Sunderland
43 Thornley
44 Tow Law
45 Tunstall
46 Usworth
47 Washington
48 West Auckland
49 Whickham
50 Whitburn
51 Willington
52 Wingate
53 Witton Gilbert

Essex (ESS)

1 Braintree
2 Brentwood
3 Buckhurst Hill
4 Brightlingsea
5 Burnham-on-Crouch
6 Chelmsford
7 Chingford
8 Clacton
9 Colchester
10 Coggeshall
11 Epping
12 Grays Thurrock
13 Halstead
14 Harwich
15 Loughton
16 Maldon
17 Romford
18 Saffron Walden
19 Shoeburyness
20 Southend-on-Sea
21 Tilbury
22 West Thurrock
23 Witham

Gloucestershire (GLO)

1 Bristol
2 Cheltenham
3 Coleford
4 Cirencester
5 Dursley
6 Gloucester
7 Nailsworth

8 Stroud
9 Tewkesbury

Hampshire (HAM)

1 Aldershot
2 Alton
3 Andover
4 Basingstoke
5 Bournemouth & Poole
6 Christchurch
7 Cowes
8 Eastleigh
9 Fareham
10 Farnborough
11 Fleet
12 Havant
13 Lymington
14 Newport
15 Petersfield
16 Portsmouth
17 Romsey
18 Ryde
19 Sandown
20 Shanklin
21 St Helens
22 Southampton
23 Totton
24 Warblington & Emsworth
25 Winchester
26 Ventnor

Herefordshire (HFD)

1 Hereford
2 Ledbury
3 Leominster
4 Ross-on-Wye

Hertfordshire (HRT)

1 Barnet
2 Berkhamsted
3 Bishop's Stortford
4 Bushey
5 Cheshunt
6 East Barnet
7 Harpenden
8 Hemel Hempstead
9 Hertford
10 Hitchin
11 Hoddesdon
12 Letchworth
13 Rickmansworth
14 Royston
15 St Albans
16 Stevenage
17 Tring
18 Ware
19 Watford

Huntingdonshire (HUN)

1 Huntingdon
2 Ramsey
3 St Ives
4 St Neots

Kent (KEN)

1 Ashford
2 Bexley
3 Broadstairs
4 Canterbury
5 Chislehurst
6 Crayford
7 Dartford
8 Deal
9 Dover
10 Erith
11 Farnborough
12 Faversham
13 Folkestone
14 Gravesend
15 Herne Bay
16 Hythe
17 Maidstone
18 Margate
19 Medway Towns
20 Orpington
21 Queenborough
22 Ramsgate
23 Royal Tunbridge Wells
24 Sandwich
25 Sevenoaks
26 Sheerness
27 Sidcup
28 Sittingbourne
29 Southborough
30 Stone
31 Swanscombe
32 Tonbridge
33 Whitstable

Lancashire (LAN)

1 Abram
2 Accrington
3 Adlington
4 Aspull
5 Ashton-in-Makerfield
6 Ashton-under-Lyne
7 Atherton
8 Bacup
9 Barrowford
10 Barrow-in-Furness
11 Blackburn
12 Blackpool
13 Blackrod
14 Bolton
15 Brierfield
16 Burnley
17 Bury
18 Carnforth
19 Chorley
20 Clayton-le-Moors
21 Clitheroe
22 Colne
23 Crompton
24 Darwen
25 Dalton-in-Furness
26 Denton
27 Eccles
28 Farnworth
29 Fleetwood
30 Formby
31 Golborne
32 Great Harwood
33 Haslingden
34 Haydock
35 Heywood
36 Hindley
37 Horwich
38 Huyton-with-Roby
39 Irlam
40 Kirkham
41 Lancaster
42 Leigh
43 Leyland
44 Littleborough
45 Little Hulton
46 Little Lever
47 Liverpool & Birkenhead
48 Longridge
49 Lytham
50 Manchester & Salford
51 Middleton
52 Milnrow
53 Morecambe
54 Mossley

55 Nelson
56 Newton
57 Oldham
58 Orrell
59 Ormskirk
60 Padiham
61 Pendlebury
62 Prescot
63 Preston
64 Radcliffe
65 Rainford
66 Ramsbottom
67 Rawtenstall
68 Rishton
69 Rochdale
70 Royton
71 St Annes on Sea
72 St Helens
73 Skelmersdale
74 Southport
75 Standish
76 Thornton
77 Tottington
78 Tyldesley
79 Upholland
80 Ulverston
81 Urmston
82 Walton-le-Dale
83 Wardle
84 Warrington
85 West Houghton
86 Whiston
87 Whitefield
88 Whitworth
89 Widnes
90 Wigan

Leicestershire (LEC)

1 Ashby de la Zouch
2 Barwell
3 Coalville
4 Earl Shilton
5 Enderby
6 Hinckley
7 Hugglescote
8 Ibstock
9 Leicester
10 Loughborough
11 Market Harborough
12 Melton Mowbray
13 Oadby
14 Shepshed
15 Sileby
16 Syston
17 Whittick
18 Wigston Magna

Lincolnshire (LIN)

1 Alford
2 Barton-on-Humber
3 Boston
4 Brigg
5 Gainsborough
6 Grantham
7 Grimsby
8 Horncastle
9 Lincoln
10 Louth
11 Market Rasen
12 Scunthorpe
13 Sleaford
14 Spalding
15 Stamford
16 Skegness

London

Middlesex (MDX)

1 Ashford
2 Enfield
3 Feltham
4 Friern Barnet
5 Hampton
6 Harrow
7 Hayes
8 Hounslow
9 Pinner
10 Southall
11 Staines
12 Sunbury-on-Thames
13 Teddington
14 Twickenham
15 Uxbridge
16 Wembley
17 Yiewsley & West Drayton

Norfolk (NFK)

1 Cromer
2 Diss
3 Downham Market
4 East Dereham
5 Fakenham
6 Great Yarmouth
7 Hunstanton
8 King's Lynn
9 North Walsham
10 Norwich
11 Thetford
12 Sheringham
13 Wells-next-the-Sea

Northamptonshire (NTS)

1 Burton Latimer
2 Daventry
3 Desborough
4 Earls Barton
5 Finedon
6 Higham Ferrers
7 Irthlingborough
8 Kettering
9 Northampton
10 Oundle
11 Peterborough
12 Raunds
13 Rothwell
14 Rushden
15 Wellingborough

Northumberland (NLD)

1 Alnwick
2 Amble
3 Ashington
4 Bedlington
5 Berwick
6 Blyth
7 Cramlington
8 East Chevington
9 Haltwhistle
10 Hexham
11 Morpeth
12 Newbiggin-by-Sea
13 Newburn
14 Newcastle & Gateshead
15 Pegswood
16 Prudhoe
17 Tynemouth
18 Wallsend
19 Whitley Bay

Nottinghamshire (NOT)

1 Arnold
2 Beeston

3 Carlton
4 East Retford
5 Eastwood
6 Hucknall
8 Kimberly
9 Kirkby in Ashfield
10 Mansfield
11 Mansfield Woodhouse
12 Newark
13 Nottingham
14 Radcliffe on Trent
15 Skegby
16 Southwell
17 Stapleford
18 Sutton in Ashfield
19 Warsop
20 Worksop

Oxfordshire (OXO)

1 Banbury
2 Bicester
3 Chipping Norton
4 Henley-on-Thames
5 Oxford
6 Thame
7 Witney

Rutland (RUT)

1 Oakham
2 Uppingham

Shropshire (SAL)

1 Bridgnorth
2 Broseley
3 Dawley
4 Ludlow
5 Market Drayton
6 Newport
7 Oakengates
8 Oswestry
9 Shrewsbury
10 Wellington
11 Whitchurch

Somerset (SOM)

1 Bath
2 Bridgwater
3 Burnham-on-Sea
4 Chard
5 Clevedon
6 Crewkerne
7 Frome
8 Glastonbury
9 Midsomer Norton
10 Minehead
11 Portishead
12 Radstock
13 Shepton Mallet
14 Street
15 Taunton
16 Wellington
17 Wells
18 Weston-super-Mare
19 Yeovil

Staffordshire (STF)

1 Amblecote
2 Biddulph
3 Bilston
4 Brierley Hill
5 Brownhills
6 Burslem
7 Burton-on-Trent
8 Cannock
9 Coseley
10 Darlaston
11 Fenton
12 Hanley
13 Kidsgrove
14 Leek
15 Lichfield
16 Longton
17 Newcastle under Lyme
18 Quarry Bank
19 Rowley Regis
20 Rugeley
21 Sedgley
22 Short Heath
23 Stafford
24 Stoke-on-Trent
25 Stone
26 Tamworth
27 Tipton
28 Tunstall
29 Uttoxeter
30 Walsall
31 Wednesbury
32 Wednesfield
33 West Bromwich
34 Willenhall
35 Wolverhampton

Suffolk (SFK)

1 Beccles
2 Bungay
3 Bury St Edmunds
4 Felixstowe
5 Hadleigh
6 Halesworth
7 Haverhill
8 Ipswich
9 Leiston-cum-Sizewell
10 Lowestoft
11 Newmarket
12 Southwold
13 Stowmarket
14 Sudbury
15 Woodbridge

Surrey (SRY)

1 Ashtead
2 Byfleet
3 Camberley
4 Caterham
5 Cheam
6 Chertsey
7 Dorking
8 Egham
9 Epsom
10 Esher
11 Ewell
12 Farnham
13 Godalming
14 Guildford
15 Haslemere
16 Kingston upon Thames
17 Leatherhead
18 Malden
19 East & West Molesey
20 Purley
21 Reigate
22 Sutton
23 Wallington
24 Walton-on-Thames
25 Warlingham
26 Weybridge
27 Woking

Sussex (SSX)

1 Arundel
2 Bexhill
3 Bognor

4 Brighton & Hove
5 Burgess Hill
6 Chichester
7 Eastbourne
8 East Grinstead
9 Hastings
10 Haywards Heath
11 Horsham
12 Lewes
13 Littlehampton
14 Newhaven
15 Rye
16 Seaford
17 Shoreham
18 Uckfield
19 Worthing

Warwickshire (WAR)

1 Atherstone
2 Bedworth
3 Birmingham & Smethwick
4 Coventry
5 Kenilworth
6 Nuneaton
7 Royal Leamington Spa
8 Rugby
9 Stratford-on-Avon
10 Sutton Coldfield
11 Warwick

Westmorland (WLD)

1 Ambleside
2 Bowness
3 Kendal

Wiltshire (WIL)

1 Bradford-on-Avon
2 Calne
3 Chippenham
4 Devizes
5 Malmesbury
6 Marlborough
7 Melksham
8 Salisbury
9 Swindon
10 Trowbridge
11 Warminster
12 Westbury

Worcestershire (WOR)

1 Bewdley
2 Bromsgrove
3 Cradley
4 Droitwich
5 Dudley
6 Evesham
7 Halesowen
8 Kidderminster
9 Lye
10 Malvern
11 Oldbury
12 Redditch
13 Stourbridge
14 Stourport
15 Worcester

Yorkshire – East Riding (YER)

1 Beverley
2 Bridlington
3 Filey
4 Gt Driffield
5 Hessle
6 Kingston upon Hull
7 Pocklington

Yorkshire – North Riding (YNR)

1 Brotton
2 Eston
3 Guiseborough
4 Loftus
5 Malton
6 Middlesbrough
7 Northallerton
8 Redcar
9 Richmond
10 Saltburn
11 Scarborough
12 Skelton
13 Thirsk
14 Whitby
15 York

Yorkshire – West Riding (YWR)

1 Adwick le Street
2 Altofts
3 Ardsley
4 Aston cum Aughton
5 Baildon
6 Barnoldswick
7 Barnsley
8 Batley
9 Bentley with Arksey
10 Bingley
11 Birkenshaw
12 Birstall
13 Bolton upon Dearne
14 Bradford
15 Brighouse
16 Burley in Wharfedale
17 Calverley
18 Castleford
19 Cleckheaton
20 Clayton
21 Conisbrough
22 Cudworth
23 Dalton
24 Darfield
25 Darton
26 Denholme
27 Dewsbury
28 Dinnington
29 Dodworth
30 Doncaster
31 Drighlington
32 Earby
33 Elland
34 Ferry Fryston
35 Featherstone
36 Farsley
37 Garforth
38 Gildersome
39 Glass Houghton
40 Glusburn
41 Goole
42 Greasbrough
43 Greetland
44 Guiseley
45 Halifax
46 Harrogate
47 Hebden Bridge
48 Heckmondwike
49 Hemsworth
50 Hipperholme
51 Honley
52 Horbury
53 Horsforth
55 Hoyland Nether
56 Huddersfield

57 Ilkley
58 Keighley
59 Kippax
60 Kirkburton
61 Kirkheaton
62 Knaresborough
63 Knottingley
64 Leeds
65 Lepton
66 Linthwaite
67 Meltham
68 Menston
69 Methley
70 Mexborough
71 Mirfield
72 Monks Bretton
73 Morley
74 Mytholmroyd
75 Normanton
76 Ossett
77 Otley
78 Penistone
79 Pontefract
80 Pudsey
81 Queensbury
82 Rawdon
83 Rawmarsh
84 Ripon
85 Rotherham
86 Rothwell
87 Royston
88 Saddleworth
89 Selby
90 Sheffield
91 Shelf
92 Shipley
93 Silsden
94 Skelmanthorpe
95 Skipton
96 Slaithwaite
97 South Elmsall
98 Southowram
99 Sowerby
100 Sowerby Bridge
102 Stainland
103 Stanley
104 Stocksbridge
105 Swinton
106 Tadcaster
108 Todmorden
109 Wakefield
110 Wales
111 West Melton
112 Wombwell
113 Woodlesford
114 Worsborough
115 Wath upon Dearne
116 Yeadon

Monmouthshire (MON)

1 Abercarn
2 Abergavenny
3 Abersychan
4 Abertillery
5 Bedwellty
6 Blaenavon
7 Chepstow
8 Ebbw Vale
9 Llanfrecha Upper
10 Monmouth
11 Mynyddislwyn
12 Nantyglo & Blaina
13 Newport
14 Panteg
15 Pontypool
16 Rhymney
17 Risca
18 Tredegar

North Wales (NWA)

1 Holyhead
2 Aberystwyth
3 Cardigan
4 Bangor
5 Bethesda
6 Caernarvon
7 Conway
8 Llandudno
9 Penmaenmawr
10 Portmadoc
11 Pwllheli
12 Colwyn Bay
13 Denbigh
14 Llangollen
15 Llanrwst
16 Ruthin
17 Wrexham
18 Buckley
19 Connah's Quay
20 Rhyl
21 Flint
22 Holywell
23 Mold
24 Llanidloes
25 Newtown
26 Llandrindod Wells

South Wales (SWA)

1 Brecon
2 Brynmawr
3 Ammanford
4 Burry Port
5 Carmarthen
6 Cwm Amman
7 Kidwelly
8 Llanelli
9 Aberdare
10 Barry
11 Bridgend
12 Briton Ferry
13 Cardiff
14 Caerphilly
15 Gelligaer
16 Maesteg
17 Merthyr Tydfil
18 Mountain Ash
19 Neath
20 Ogmore
21 Penarth
22 Pontypridd
23 Porthcawl
24 Port Talbot
25 Rhondda
26 Swansea
27 Fishguard
28 Haverfordwest
29 Milford Haven
30 Neyland
31 Pembroke
32 Tenby

Appendix B Reference map overleaf

Appendix B Reference map overleaf

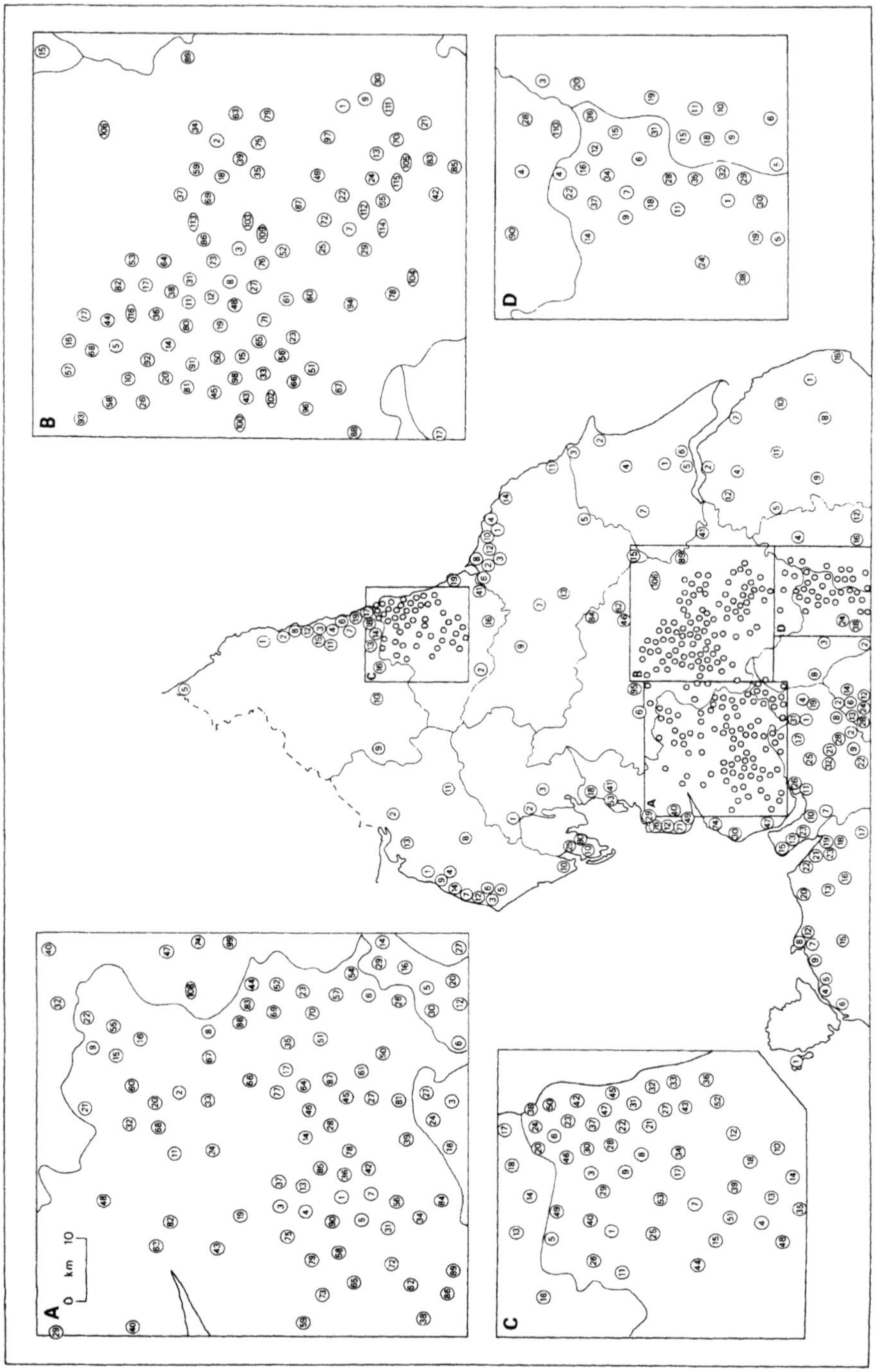
A
0 km 10
B
C
D

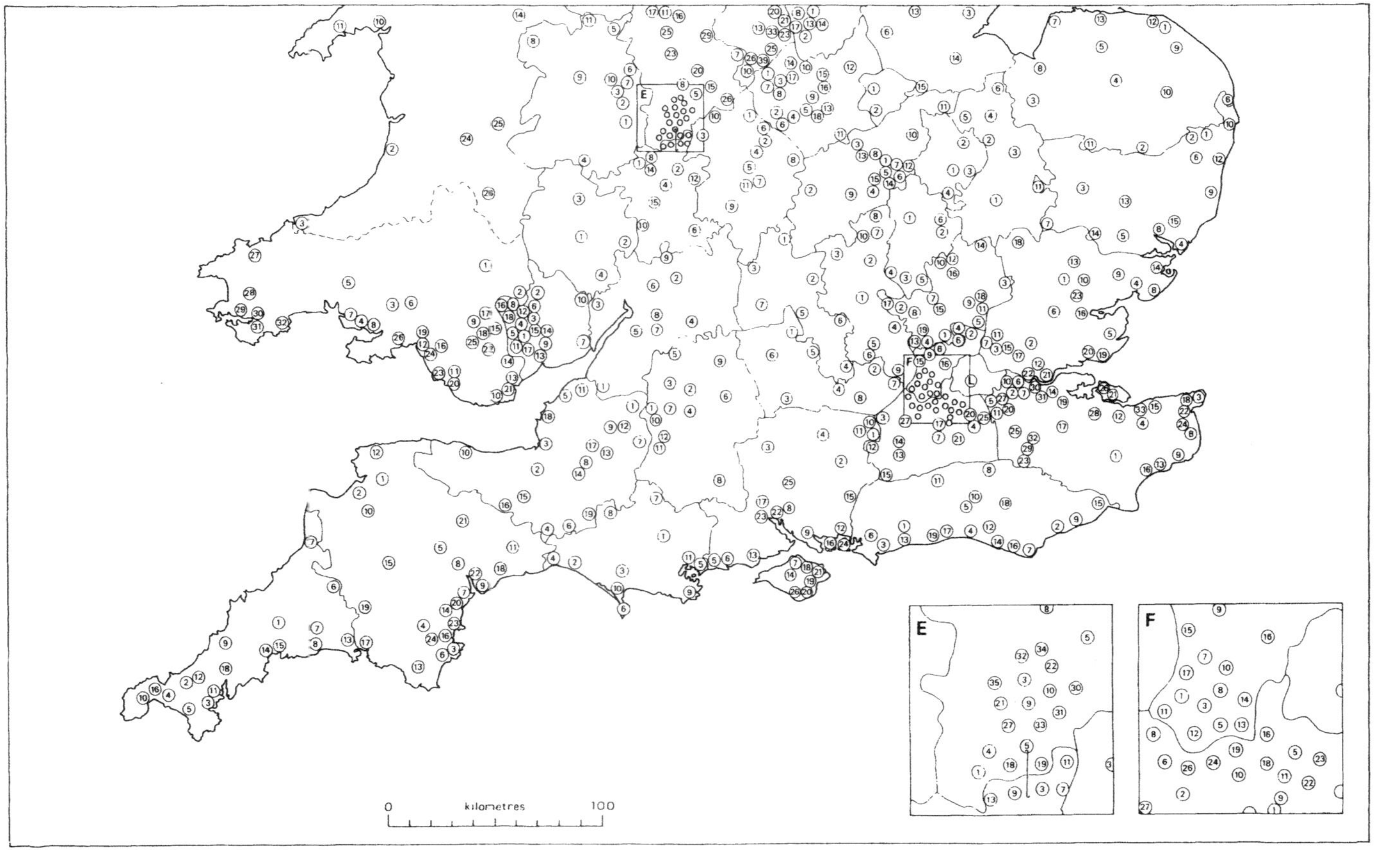

Appendix B Reference map: towns used in the analysis

Sources

The following is a list of sources rather than a bibliography, since, in covering a topic which straddles so many fields, a comprehensive bibliography would be impossible. Many primary sources were consulted, especially in connection with the material on urban innovations. The assistance given by numerous librarians and keepers of archives – especially those of the Post Office archives – was of great value and I should like to record my appreciation of their help.

Of the primary sources consulted – in addition to the *Reports* of the Census for England and Wales – the following were most extensively relied upon:

(i) PARLIAMENTARY PAPERS

For the regulation of Benefit Building Societies, 1836.I.

Return relating to the condition of Benefit Societies, etc. 1859.XXIII.

Return(s) of Societies Incorporated under the Building Society Acts, etc: 1876.LXVIII; 1877.LXXVI; 1878.LXVIII; 1878-9.LXV; 1880.LXVII; 1881.LXXXIII, 1882.LXIV; 1883.LXIV; 1884.LXXII; 1884-5.LXXII; 1886.LXII; 1887.LXXVIII, 1888.XCIII; 1889.LXXII; 1890-1.LXXIX.

Return of Building Societies . . . which have terminated, etc. 1892.LXXIII.

Tables showing Number of Building Societies . . . Registered . . . in each year 1874-91, etc. 1893-4.LXXXIV.

Reports to Government respecting Gas-Light Establishments. 1823.V.

Statement from every Gas Company established by Act of Parliament etc. 1874.XLIV

Abstract of Return of Gas Companies, etc. 1850.XLIX.

Return from every Gas Company, etc: 1865.L; 1866.LXVI.

(ii) DIRECTORIES

The Building and Freehold Land Societies' Directory and Almanack for 1853, etc. Rouse & Co. (London) 1853.

The Building Societies' Directory and Almanack for 1854; idem, 1855, Edited by Henry Brooks (London) 1854, 1855.

Directory and Handbook of Building and Freehold Land Societies etc. Building Societies' Gazette (London) 1874.

The Directory of Building Societies, etc. Compiled by H. Kent and V. M. Braund (London) 1884, etc.

Gas and Water (and electric lighting) Companies' Directory (London) 1877-94.

Gas and electric lighting companies' (works) directory and statistics (London) 1895-9.

Gas works directory and statistics (London) 1900-30.

Electrician's Directory (London) 1883-5.

The Electrician: electrical trades' directory and handbook (London) 1894-1918.

Manual of electrical undertakings etc. Compiled under the direction of E. Garcke (London) 1896-1947.

(iii) TELEPHONE HISTORY

Material on the dates of opening of exchanges was drawn largely from the periodical *The Electrician*, from early telephone directories (see above, p. 169*n*) and from local newspapers. In addition, the records of the Post Office archives, Post 30, proved invaluable for the wider history of early telephone development.

References

Ackerman, E. A. (1963) Where is a research frontier?, *Ann. Ass. Am. Geogr.* 53, 429–40.

Adelman, I. G. (1958) A stochastic model of the size distribution of firms, *J. Am. Statist. Ass.* 53, 893-904.

Ajo, R. (1969) Response of geographical systems to repeated psychological shock, *Geogr. Annlr, Ser. B.* 51, 8–14.

Aldcroft, D. H. (1964) The entrepreneur and the British economy, 1870–1913, *Econ. Hist. Rev.* 2nd Ser. 17, 113–34.

Aldcroft, D. H. (1966) Technical progress and British enterprise, 1875–1914, *Busin. Hist.* 8, 122–39.

Aldcroft, D. H. (ed.) (1968) *The development of British industry and foreign competition, 1875–1914: case studies of ten industries* (London).

Allen, G. R. (1954) The 'courbes des populations': a further analysis, *Bull. Oxf. Univ. Inst. Statist.* 16, 179–89.

Angyal, A. (1941) *Foundations for a science of personality* (Cambridge, Mass.).

Anthony, J. and Baxter, R. (1971) The first stage in disaggregating the residential sub-model, *Working Pap.* No. 58, Land Use and Built Form Studies, Univ. of Cambridge.

Armstrong, W. (1972) The use of information about occupation, in Wrigley, E. A. (ed.), *Nineteenth-century society: essays in the use of quantitative methods* (Cambridge), pp. 191–310.

Ashworth, W. (1960) *An economic history of England, 1870–1939* (London).

Ashworth, W. (1966) The late Victorian economy, *Economica* 33, 17–33.

Aspin, A. C. and Welch, B. L. (1949) Tables for use in comparisons whose accuracy involves two variances, separately estimated, *Biometrika* 36, *290*–6.

Association of American Geographers (1968) A systems analytic approach to economic geography, *Publication*, No. 8, Commission on College Geography Publications, Ass. Am. Geogr.

Auerbach, F. (1913) Das Gesetz der Bevölkerungskonzentration, *Petermanns Mitt.* 59, No. 1, 74–6.

Bailey, N. T. J. (1957) *The mathematical theory of epidemics* (New York).

Baldwin, F. G. C. (1925) *History of the telephone in the United Kingdom* (London).

Banks, J. A. (1968) Population change and the Victorian city, *Victorian Stud.* 11, 277–89.

Bassett, K. A. and Haggett, P. (1971) Towards short-term forecasting for cyclic behaviour in a regional system of cities, in Chisholm, M. *et al.* (eds), *Regional forecasting* (London), pp. 389-413.

Beckmann, M. J. (1958) City hierarchies and the distribution of city size, *Econ. Dev. cult. Change* 6, 243–8.

Beckmann, M. J. and McPherson, J. C. (1970) City size distribution in a central place hierarchy: an alternative approach, *J. reg. Sci.* 10, 25–33.

Bell, G. (1962) Change in city size distribution in Israel, *Ekistics*, 13, No. 75, 103.

Bellamy, J. (1952) Occupations in Kingston upon Hull, 1841–1948, *Yorks. Bull. econ. soc. Res.* 4, 33–50.

Beresford, M. W. (1971) The back-to-back house in Leeds, 1787–1937, in Chapman, S. D. (ed.) (1971), pp. 93–132.

Berry, B. J. L. (1961) City size distribution and economic development, *Econ. Dev. cult. Change* 9, 573–87.

Berry, B. J. L. (1964) Cities as systems within systems of cities, *Pap. Reg. Sci. Ass.* 13, 147–63.

Berry, B. J. L. (1967) *Geography of market centres and retail distribution* (Englewood Cliffs, N J).

Berry, B. J. L. (1968) Metropolitan Area redefinition: a re-evaluation of concept and statistical practice, *Working Pap.* No. 28, US Bureau of the Census, Washington DC.

Berry, B. J. L. (1969) City size and economic development: conceptual synthesis and policy problems, with special reference to South and South-East Asia, *Unpublished ms.* Dept. of Geogr., Univ. of Chicago.

Berry, B. J. L. (1970) The geography of the United States in the year 2000, *Trans. Inst. Br. Geogr.* 51, 21–53.

Berry, B. J. L. (1972) Hierarchical diffusion: the basis of development filtering and spread in a system of growth centres, in English, P. W. and Mayfield, R. C. (eds), *Man, space and environment: concepts in contemporary human geography* (London), pp. 340–59.

Berry, B. J. L. and Garrison, W. L. (1958) Alternate explanations of urban rank-size relationships, *Ann. Ass. Am. Geogr.* 48, 83–91.

Berry, B. J. L. and Horton, F. E. (1970) *Geographic perspectives on urban systems* (Englewood Cliffs, N J).

Berry, B. J. L. and Pred, A. R. (1965) Central place studies: a bibliography of theory and applications including supplement through 1964, *Biblphy Ser.* No. 1, Reg. Sci. Res. Inst., Philadelphia.

Beshers, J. M. and Laumann, E. O. (1967) Social distance: a network approach, *Am. sociol. Rev.* 32, 225–36.

Best, G. (1971) *Mid-Victorian Britain, 1851–75* (London).

Boal, F. W. (1968) Technology and urban form, *J. Geogr.* 67, 229–36.

Bogue, D. J. (1950) *The structure of the metropolitan community* (Ann Arbor, Mich.).

Boon, F. (1967) A simple model for the diffusion of an innovation in an urban system, *Reg. Econ. Dev. Stud.* No. 1, Center for Urban Studies, Univ. of Chicago.

Booth, C. (1886) Occupations of the people of the United Kingdom, 1801–81, *Jl R. Statist. Soc.* 49, 314-435.

Booth, C. (1902–3) *Life and labour of the people in London, etc.* (London), 17 vols.

Borchert, J. H. (1967) American metropolitan evolution, *Geogrl Rev.* 57, 301-31.

Boulding, K. E. (1956) General System Theory: the skeleton of science, *Mgmt Sci.* 2, 197–208.

Bowers, R. V. (1937) The direction of intra-societal diffusion, *Am. sociol. Rev.* 2, 826–36.

Bowkett, T. E. (1844) *Alchemy, or, the art of converting the baser metals into gold . . . a highly valuable receipt for the manufacture of an accumulating fund, investment, provident, building, equitable association* (London).

Bowkett, T. E. (1850) *The bane and the antidote, etc.* (London).

Briggs, A. (1963) *Victorian cities* (London).

Briggs, A. (1968) The Victorian city: quantity and quality, *Victorian Stud.* 11, 711–30.

Brooks, E. (1967) Public housing and labour mobility in Britain, *Unpublished Pap.* Study Group in Urban Geogr., Inst. Br. Geogr.

Brown, L. A. (1968) Diffusion processes and location: a conceptual framework and bibliography, *Biblphy Ser.* No. 4, Reg. Sci. Res. Inst., Philadelphia.

Brown, L. A. and Moore, E. G. (1969) Diffusion research in geography: a perspective, in Board, C. *et al.* (eds), *Progress in Geography*, Vol. 1 (London), pp. 119-57.

Browning, H. L. and Gibbs, J. P. (1961) Some measures of demographic and spatial relationships among cities, in Gibbs, J. P. (ed.) (1961), pp. 436–59.

Burn, D. L. (1940) *The economic history of steelmaking, 1867-1939* (Cambridge).

Butlin, N. G. (1964) *Investment in Australian economic development, 1861–1900* (Cambridge).

Byatt, I. C. R. (1968) Electrical products, in Aldcroft, D. H. (ed.) (1968), pp. 238–73.

Cairncross, A. K. (1949) Internal migration in Victorian England, *Manchr Sch.* 17, 67–87.

Cairncross, A. K. (1953) *Home and foreign investment, 1870–1913: studies in capital accumulation* (Cambridge).

Carr-Saunders, A. M. and Jones, D. C. (1927) *A survey of the social structure of England and Wales as illustrated by statistics* (Oxford).

Carter, H. (1955) Urban grades and spheres of influence in south-west Wales: an historical consideration, *Scott. Geogr. Mag.* 71, 43–58.

Carter, H. (1969) The growth of the Welsh city system, *Inaugural lecture*, University College of Wales, Aberystwyth (Cardiff).

Carter, H. (1972) *The study of urban geography* (London).

Casetti, E. (1967) Urban population density patterns: an alternate explanation, *Can. Geogr.* 11, 96–100.

Casetti, E. and Semple, R. K. (1968) A method for the stepwise separation of spacial trends, *Mich. Int-Com. Math. Geogr.* No. 11.

Casetti, E. and Semple, R. K. (1969) Concerning the testing of spatial diffusion hypotheses, *Geogrl Analysis* 1, 254–9.

Census (1923) *Census of England and Wales, 1911. Fertility of marriage,* Part 2 (London).

Champernowne, D. G. (1937) The distribution of income between persons, *Unpublished Fellowship Dissertation*, Vol. 2, King's College, Cambridge.

Chapman, G. P. (1970) The application of information theory to the analysis of population distribution in space, *Econ. Geogr.* 46, 317–31.

Chapman, G. P. (1971) The object of geographical analysis, *Unpublished Pap.* Comn Quant. Meth., Int. Geogrl Un., Budapest conference.

Chapman, S. D. (ed) (1971) *The history of working-class housing: a symposium* (Newton Abbot).

Childe, V. G. (1936) *Man makes himself* (London).

Chisholm, G. G. (1897) On the distribution of towns and villages in England, Part 2: Historical aspects, *Geogrl J.* 10, 511–30.

Chisholm, M. (1967) General Systems Theory and geography, *Trans. Inst. Br. Geogr.* 42, 45–52.

Chorley, R. J. and Haggett, P. (1965) Trend-surface mapping in geographical research, *Trans. Inst. Br. Geogr.* 37, 47–67.

Chorley, R. J. and Kennedy, B. A. (1971) *Physical geography: a systems approach* (London).

Christaller, W. (1933) *Central places in southern Germany* (original publication, 1933: English translation by C. W. Baskin, 1966) (Englewood Cliffs, NJ).

Church, R. A. (1966) *Economic and social change in a Midland town: Victorian Nottingham, 1815–1900* (London).

Clark, C. (1951) Urban population densities, *Jl R. Statist. Soc., Ser. A.* 114, 490–6.

Clark, C. (1967) *Population growth and land use* (London).

Clark, G. Kitson (1962) *The making of Victorian England* (London).

Cleary, E. J. (1965) *The building society movement* (London).

Clegg, S. (1841) *A practical treatise on the manufacture and distribution of coal-gas* (London).

Cliff, A. D. (1968) The neighbourhood effect in the diffusion of innovations, *Trans. Inst. Br. Geogr.* 44, 75–84.

Cliff, A. D. and Ord, K. (1970) Spatial autocorrelation: a review of existing and new measures with applications, *Econ. Geogr.* 46, 269–92.

Crain, R. L. (1966) Fluoridation: the diffusion of an innovation among cities, *Social Forces* 44, 467–76.

Cripps, E. L. and Foot, D. H. S. (1969) The empirical development of a residential location model for use in sub-regional planning, *Envirmt Plann.* 1, 81-90.

Crowther, D. and Echenique, M. (1972) Development of a model of spatial structure, in Martin, L. and March, L. (eds), *Urban space and structures* (Cambridge), pp. 175–218.

Curry, L. (1964) The random spatial economy: an exploration in settlement theory, *Ann. Ass. Am. Geogr.* 54, 138–46.

Curry, L. (1967) Central places in the random spatial economy, *J. reg. Sci.* 7, No. 2, 217–38.

Dacey, M. F. (1966) Population of places in a central place hierarchy, *J. reg. Sci.* 6, No. 2, 27–33.

Dacey, M. F. (1968) An empirical study of the areal distribution of houses in Puerto Rico, *Trans. Inst. Br. Geogr.* 45, 51–69.

D'Arcy Thompson, W. (1917) *On growth and form* (Cambridge) (abridged ed., edited by J. T. Bonner, 1961).

Davies, A. (1926) Critical analysis of the census returns of the Merthyr Tydfil area, *Geogrl Teacher* 13, 473–9.

Deane, P. and Cole, W. A. (1962) *British economic growth, 1688–1959: trends and structure* (Cambridge).

Dickinson, R. E. (1947) *City, region and regionalism: a geographical contribution to human ecology* (London).

Donnison, D. V. (1967) *The government of housing* (London).

Drake, M. (1972) The Census, 1801–1891, in Wrigley, E. A. (ed.), *Nineteenth-century society: essays in the use of quantitative methods* (Cambridge), pp. 7–46.

Du Boff, R. B. (1967) The introduction of electric power in American manufacturing, *Econ. Hist. Rev.* 2nd Ser. 20, 509–18.

Duncan, O. D. and Reiss, A. J. (1956) *Social characteristics of urban and rural communities, 1950* (New York).

Duncan, O. D. *et al.* (1960) *Metropolis and region* (Baltimore).

Dyos, H. J. (1953) Workman's fares in south London, 1860–1914, *J. Transp. Hist.* 1, 3–19.

Dyos, H. J. (1955) Railways and housing in Victorian London, *J. Transp. Hist.* 2, 11–21, 90–100.

Evans, A. W. (1972) The pure theory of city size in an industrial economy, *Urban Stud.* 9, 49–77.

Falkus, M. E. (1967) The British gas industry before 1850, *Econ. Hist. Rev.* 2nd Ser. 20, 494–508.

Feller, I. (1966) The draper loom in New England textiles, 1894–1914: a study of diffusion of an innovation, *J. econ. Hist.* 26, 320–47.

Feller, I. (1971) The urban location of U.S. invention, 1860–1910, *Explor. entrepreneurial Hist.* 8, 285–304.

Ferguson, R. I. (1969) The rank-size rule. Parts 1 and 2, *Geogrl Artic.* No. 12, 36–42 and No. 13, 25–44, Dept. of Geogr., Univ. of Cambridge.

Fiebleman, J. K. (1954) Theory of integrative levels, *Br. J. Philos. Sci.* 5, 59–90.

Fisher, F. J. (1935) The development of the London food market, 1540–1640, *Econ. Hist. Rev.* 5, 46–64.

Frankel, M. (1955) Obsolescence and technological change in a maturing economy, *Am. econ. Rev.* 45, 296–319.

Friedlander, D. (1970) The spread of urbanization in England and Wales, 1851–1951, *Popul. Stud.* 24, 423–43.

Friedlander, D. and Roshier, R. J. (1966) A study of internal migration in England and Wales, Part 1: geographical patterns of internal migration, *Popul. Stud.* 19, 239–78.

Friedmann, J. (1966a) *Regional development policy: a case study in Venezuela* (Cambridge, Mass.).

Friedmann, J. (1966b) Two concepts of urbanization: a comment, *Urban Affairs* 1, No. 4, 79–84.

Friedmann, J. and Miller, J. (1965) The urban field, *J. Am. Inst. Planners* 31, 312–20.

Fuguitt, G. W. (1965) The growth and decline of small towns as a probability process, *Am. sociol. Rev.* 30, 403–11.

Garke, E. (1907) *The progress of electrical enterprise* (London).

Geddes, P. E. (1949) *Cities in evolution* (London, revised ed.).

Gibbs, J. P. (ed) (1961) *Urban research methods* (New Jersey).

Gibrat, R. (1931) *Les inégalités économiques* (Paris).

Goheen, P. G. (1970) Victorian Toronto, 1850–1900: pattern and process of growth, *Res. Pap.* No. 127, Dept. of Geogr., Univ. of Chicago.

Goodrich, E. P. (1926) The statistical relationship between population and the city plan, in Burgess, E. W. (ed.), *The urban community* (Chicago), pp. 144–50.

Gottmann, J. (1961) *Megalopolis: the urbanized northeastern seabord of the United States* (New York).

Granger, C. W. J. (1964) *Spectral analysis of economic time series* (Princeton, N J).

Gras, N. S. B. (1922) *An introduction to economic history* (New York).

Gunston, W. H. (1907–8) The development of the telephone in the English counties since 1892 geographically considered, *Natn. Teleph. J.* 2, 134–5, 158–9, 183–5, 228–30.

Gunston, W. H. (1910) The National Telephone Company in country districts, *Natn. Teleph. J.* 6, 73–4.

Habakkuk, H. J. (1962a) *American and British technology in the nineteenth century: the search for labour-saving inventions* (Cambridge).

Habakkuk, H. J. (1962b) Fluctuations in house-building in Britain and the United States in the nineteenth century, *J. econ. Hist.* 22, 198–230.

Hägerstrand, T. (1966) Aspects of the spatial structure of social communications and the diffusion of innovations, *Pap. Reg. Sci. Ass.* 16, 27–42.

Hägerstrand, T. (1967a) *Innovation diffusion as a spatial process* (Chicago, English translation by A. R. Pred) (original Swedish edition, 1953).

Hägerstrand, T. (1967b) On Monte Carlo simulation of diffusion, in Garrison, W. L. and Marble, D. F. (eds), Quantitative geography, Part 1: economic and cultural topics, *NWest. Univ. Stud. Geogr.* No. 13, 1-32.

Hall, A. D. and Fagen, R. E. (1956) Definition of system, *Gen. Syst.* 1, 18–28.

Hall, P. G. (1962) *The industries of London since 1861* (London).

Hall, P. G. (1966) *The world cities* (London).

Harris, C. D. (1970) *Cities of the Soviet Union: studies in their functions, size, density and growth* (Chicago).

Hartwell, R. M. (1971) *The industrial revolution and economic growth* (London).

Harvey, D. (1966) Geographical processes and point patterns: testing models of diffusion by quadrat methods, *Trans. Inst. Br. Geogr.* 40, 81–95.

Harvey, D. (1968) Some methodological problems in the use of Neyman Type A and negative binomial probability distributions in the analysis of spatial series, *Trans. Inst. Br. Geogr.* 44, 85–95.

Harvey, D. (1969) *Explanation in geography* (London).

Higgs, R. (1969) The growth of cities in a midwestern region, 1870–1900, *J. reg. Sci.* 9, 369–75.

Hirschman, A. O. (1958) *The strategy of economic development* (New Haven, Conn.).

Hobsbawm, E. J. (1964) The nineteenth century London labour market, in Centre for Urban Studies, *London: aspects of change* (London), pp. 3–28.

Hudson, J. C. (1969) Diffusion in a central place system, *Geogrl Analysis* 1, 45–58.

Hughes, T. P. (1962) British electrical industry lag, 1882–1888, *Technology Cult.* 3, 27–44.

Ijiri, H. and Simon, H. A. (1964) Business firm growth and size, *Am. econ. Rev.* 54, 77–89.

Innes, J. W. (1938) *Class fertility trends in England and Wales, 1876–1934* (Cambridge).

Interdepartmental Committee on Social and Economic Research (1951) Census reports of Great Britain, 1801–1931, *Guides to Official Sources*, No. 2 (London).

International Urban Research (1959) *The world's metropolitan areas* (Berkeley, Calif.).

Jefferson, M. (1939) The law of the primate city, *Geogrl Rev.* 29, 226–32.

Jones, G. S. (1971) *Outcast London: a study in the relationship between classes in Victorian society* (Oxford).

Kalecki, M. (1945) On the Gibrat distribution, *Econometrica* 13, 161–70.

Keeble, D. E. (1967) Models of economic development, in Chorley, R. J. and Haggett, P. (eds), *Models in geography* (London), pp. 243–302.

Kellett, J. R. (1969) *The impact of railways on Victorian cities* (London).

Kemeney, J. G. and Snell, J. L. (1960) *Finite Markov chains* (Princeton, NJ).

Kendall, M. G. (1961) Natural law in the social sciences, *Jl R. Statist. Soc., Ser. A.* 124, 2–16.

King, L. J. E., Casetti, E. and Jeffrey, D. (1969) Economic impulses in a regional system of cities, *Reg. Stud.* 3, 213–18.

Kingsbury, J. E. (1915) *The telegraph and telephone exchanges: their invention and development* (London).

Krause, J. T. (1958) Changes in English fertility and mortality, 1781–1850, *Econ. Hist. Rev.* 2nd Ser. 11, 52–70.

Krim, A. J. (1967) The innovation and diffusion of the street railway in North America, *Unpublished MA Dissertation,* Univ. of Chicago.

Lampard, E. E. (1968) The evolving system of cities in the United States: urbanization and economic development, in Perloff, H. S. and Wingo, L. (eds), *Issues in urban economics* (Baltimore), pp. 81–139.

Landes, D. S. (1969) *The unbound Prometheus: technological change and industrial development in Western Europe from 1750 to the present* (Cambridge).

Langton, J. (1972) Potentialities and problems of adopting a systems approach to the study of change in human geography, in Board, C. *et al.* (eds), *Progress in Geography*, Vol. 4 (London), pp. 125–79.

Law, C. M. (1967) The growth of urban population in England and Wales, 1801–1911, *Trans. Inst. Br. Geogr.* 41, 125–43.

Lawton, R. (1968a) The journey-to-work in Britain: some trends and problems, *Reg. Stud.* 2, 27–40.

Lawton, R. (1968b) Population changes in England and Wales in the later nineteenth century: an analysis of trends by Registration Districts, *Trans. Inst. Br. Geogr.* 44, 55–74.

Leech, B. T. (1897–8) Tramways and their municipalisation, *Trans. Manchr Statist. Soc.* 129–52.

Lever, W. F. (1972) A Markov approach to the optimal size of cities in England and Wales, *Unpublished Pap.* Quant. Meth. Stud. Group, Inst. Br. Geogr., Sheffield conference.

Levine, A. L. (1967) *Industrial retardation in Britain, 1880–1914* (New York).

Lewis, J. Parry (1965) *Building cycles and Britain's growth* (London).

Lewis, S. (1944) *A topographical dictionary of England, etc.* (London 4 vols. 5th ed.).

Lewis, W. A. (1957) International competition in manufactures, *Am. econ. Rev.* 47, No. 2, 578–87.

Liepmann, K. K. (1944) *The journey to work: its significance for industrial and community life* (London).

Loewenstein, L. K. (1965) *The location of residences and work places in urban areas* (New York).

Lopes, A. S. (1972) The economic functions of small towns and rural centres, *Unpublished PhD Dissertation*, Univ. of Oxford.

Lord, J. (1923) *Capital and steam-power, 1750–1800* (London).

Lotka, A. J. (1924) *Elements of physical biology* (Baltimore) (new ed., New York, 1965, entitled *Elements of mathematical biology*).

Lotka, A. J. (1926) The frequency distribution of scientific productivity, *J. Wash. Acad. Sci.* 16, 317–23.

Lubenow, W. C. (1971) *The politics of government growth: early Victorian attitudes towards state intervention* (Newton Abbot).

Luckermann, F. (1966) Empirical expressions of nodality and hierarchy in a circulation manifold, *E. Lakes Geogr.* 2, 17–44.

McCloskey, D. N. (ed.) (1971) *Essays on a mature economy: Britain after 1840* (London).

McCloskey, D. N. and Sandberg, L. G. (1971) From damnation to redemption: judgements on the Late Victorian entrepreneur, *Explor. entrepreneurial Hist.* 9, 89–108.

McElrath, D. C. (1968) Societal scale and social differentiation: Accra, Ghana, in Greer, S. *et al.* (eds), *The new urbanization* (New York), pp. 33–52.

McKenzie, R. D. (1933) *The metropolitan community* (New York).

Mackinder, H. J. (1902) *Britain and the British seas* (London).

McVoy, E. C. (1940) Patterns of diffusion in the United States, *Am. sociol. Rev.* 5, 219–27.

Madden, C. H. (1956a) On some indicators of stability in the growth of cities in the United States, *Econ. Dev. cult. Change* 4, 236-52.

Madden, C. H. (1956b) Some spatial aspects of urban growth in the United States, *Econ. Dev. cult. Change* 4, 371-87.

Madden, C. H. (1958) Some temporal aspects of the growth of cities in the United States, *Econ. Dev. cult. Change* 6, 143–69.

Maine, H. J. S. (1861) *Ancient Law* (London).

Maitland, F. W. (1898) *Township and borough: together with an appendix of notes relating . . . to Cambridge* (Cambridge).

Mandelbrot, B. (1953) An informational theory of the statistical structure of language, in Jackson, W. (ed.), *Communication theory* (London), pp. 486-502.

Mansfield, E. (1969) *The economics of technological change* (London) (first published, New York, 1968).

Mantoux, P. (1928) *The industrial revolution in the eighteenth century: an outline of the beginnings of the modern factory system* (English translation by Marjorie Vernon) (London).

Marsh, D. C. (1958) *The changing social structure in England and Wales, 1871–1951* (London).

Marshall, J. D. (1968) Colonisation as a factor in the planting of towns in north-west England, in Dyos, H. J. (ed), *The study of urban history* (London), pp. 215–30.

Meier, R. L. (1962) *A communication theory of urban growth* (Cambridge, Mass.).

Mesarovic, M. D. (ed.) (1964) *Views on general systems theory: proceedings of the second systems symposium at Case Institute of Technology* (New York).

Meyer, H. R. (1906) *Municipal ownership in Great Britain* (New York).

Meyer, H. R. (1907) *Public ownership and the telephone in Great Britain* (New York).

Minchinton, W. E. (1956) The diffusion of tinplate manufacture, *Econ. Hist. Rev.* 9, 349–58.

Mitchell, B. R. (1962) *Abstract of British historical statistics* (Cambridge).

Moses, L. and Williamson, H. F. (1967) The location of economic activity in cities, *Am. econ. Rev.* 57, 211–22.

Mumford, L. (1938) *The culture of cities* (London).

Mumford, L. (1961) *The city in history: its origins, its transformations, and its prospects* (London).

Myrdal, G. M. (1957) *Economic theory and under-developed regions* (London).

National Academy of Sciences (1965) The science of geography, *Report*, Natn. Res. Coun.-*ad hoc* Comm. on Geogr. (Washington).

Neft, D. S. (1966) Statistical analysis for areal distributions, *Monogr. Ser.* No. 2, Reg. Sci. Res. Inst., Philadelphia.

Newling, B. E. (1966) Urban growth and spatial structure: mathematical models and empirical evidence, *Geogrl Rev.* 56, 213–25.

Newling, B. E. (1969) The spatial variation of urban population densities, *Geogrl Rev.* 59, 242–52.

Newton, M. P. and Jeffery, J. R. (1951) *Internal migration: some aspects of population movements within England and Wales* (London).

Nordbeck, S. (1965) The law of allometric growth, *Mich. Int-Com. Math. Geogr.* No. 7.

Nordbeck, S. (1971) Urban allometric growth, *Geogr. Annlr, Ser. B.* 53, 54–67.

Olsson, G. (1965) Distance and human interaction: a review and bibliography, *Biblphy Ser.* No. 2, Reg. Sci. Res. Inst., Philadelphia.

Olsson, G. (1967) Central place systems, spatial interaction, and stochastic processes, *Pap. Reg. Sci. Ass.* 18, 13–45.

Osborne, R. H. (1964) Changes in the regional distribution of population in Great Britain, *Geography* 49, 266–73.

Parr, J. B. (1969) City hierarchies and the distribution of city sizes: a reassessment of Beckmann's contribution, *J. reg. Sci.* 9, 239–54.

Parr, J. B. (1970) Models of city size in an urban system, *Pap. Reg. Sci. Ass.* 25, 221–53.

Payne, P. L. (1967) The emergence of the large-scale company in Great Britain, 1870–1914, *Econ. Hist. Rev.* 2nd Ser. 20, 519–42.

Pearson, E. S. and Hartley, H. O. (1966) *Biometrika tables for statisticians*, vol 1 (Cambridge, 3rd ed.).

Pederson, P. O. (1970) Innovation diffusion within and between national urban systems, *Geogrl Analysis* 2, 203–54.

Perloff, H. S. *et al.* (1960) *Regions, resources and economic growth* (Baltimore).

Pirenne, H. (1925) *Medieval cities: their origins and the revival of trade* (Princeton).

Pollins, H. (1964) Transport lines and social divisions, in Centre for Urban Studies, *London: aspects of change* (London), pp. 29–61.

Popenhoe, D. (1965) On the meaning of 'urban' in urban studies, *Urban Affairs* 1, No. 1, 17–33.

Prais, S. J. (1955) Measuring social mobility, *Jl R. Statist. Soc., Ser. A.* 118, 56–66.

Prais, S. J. and Hart, P. E. (1956) The analysis of business concentration: a statistical approach, *Jl R. Statist. Soc., Ser. A.* 119, 150–75.

Pred, A. R. (1966) *The spatial dynamics of U.S. urban-industrial growth, 1800–1914: interpretive and theoretical essays* (Cambridge, Mass.).

Preece, W. H. (1882) Recent progress in telephony, *J. Soc. Arts* 30, 965–9.

Price, S. J. (1958) *Building societies: their origin and history* (London).

Price-Williams, R. (1880) On the increase of population in England and Wales, *Jl R. Statist. Soc.* 43, 462–96.

Pritchard, R. M. (1972) Intra-urban migration in Leicester, 1860–1965. *Unpublished Ph.D. Dissertation*, Univ. of Cambridge.

Quandt, R. E. (1964) Statistical discrimination among alternative hypotheses and some economic regularities, *J. reg. Sci.* 5, No. 2, 1–23.

Read, D. (1964) *The English provinces: a study in influence* (London).

Richardson, H. W. (1965a) Retardation in Britain's industrial growth, 1870–1913, *Scott. J. political Econ.* 12, 125–49.

Richardson, H. W. (1965b) Overcommitment in Britain before 1930, *Oxf. econ. Pap.* New Ser. 17, 237–59.

Richardson, H. W. (1968) Chemicals, in Aldcroft, D. H. (ed.) (1968), pp. 274–306.

Rodda, J. C. (1970) A trend-surface analysis for the planation surfaces of North Wales, *Trans. Inst. Br. Geogr.* 50, 107–14.

Rogers, E. M. (1962) *Diffusion of innovations* (New York).

Rogers, E. M. and Shoemaker, F. F. (1971) *Communication of innovations: a cross-cultural approach* (New York).

Rosing, K. E. (1966) A rejection of the Zipf model (rank size rule) in relation to city size, *Prof. Geogr.* 18, 75–82.

Ryan, B. and Gross, N. C. (1943) The diffusion of hybrid seed corn in two Iowa communities, *Rur. Sociol.* 8, 15–24.

Saul, S. B. (1962a) House building in England, 1890–1914, *Econ. Hist. Rev.* 2nd Ser. 15, 119–37.

Saul, S. B. (1962b) The motor industry in Britain to 1914, *Busin. Hist.* 5, 22–44.

Saul, S. B. (1967) The market and the development of the mechanical engineering industries in Britain, 1860–1914, *Econ. Hist. Rev.* 2nd Ser. 20, 111–30.

Saul, S. B. (1968) The machine tool industry in Britain to 1914, *Busin. Hist.* 10, 22–43.

Saville, J. (1963) Internal migration in England and Wales during the past hundred years, in Sutter, J. (ed.), *Human displacements: measurement, methodological aspects* (Paris), pp. 1–21.

Schumpeter, J. A. (1939) *Business cycles: a theoretical, historical and statistical account of the capitalist process* (New York, 2 vols.).

Scratchley, A. (1867) *A treatise on benefit building societies* (London, First ed., 1849).

Shadwell, A. (1909) *Industrial efficiency: a comparative study of industrial life in England, Germany and America* (London, First ed., 2 vols, 1906).

Shannon, H. A. (1935) Migration and the growth of London, 1841-91: a statistical note, *Econ. Hist. Rev.* 5, 79–86.

Shannon, H. A. and Grebenik, E. (1943) The population of Bristol, *Occasional Pap.* No. 2, Natn. Inst. Econ. Soc. Res., London.

Shevky, E. and Bell, W. (1955) *Social area analysis: illustrative application and computational procedure* (Stanford, Calif.).

Simon, H. A. (1955) On a class of skew distribution functions, *Biometrika* 42, 425–40.

Simon, H. A. and Bonini, C. P. (1958) The size distribution of business firms, *Am. econ. Rev.* 48, 607–17.

Singer, H. W. (1936) Courbes des populations: a parallel to Pareto's law, *Econ. J.* 46, 254–63.

Singh, A. and Whittington, G. (1968) Growth, profitability and valuation, *Occasional Pap.* No. 7, Dept. of Applied Econ., Univ. of Cambridge.

Sjoberg, G. (1960) *The preindustrial city: past and present* (New York).

Sjoberg, G. (1965) Cities in developing and in industrial societies, a cross-cultural analysis, in Hauser, P. M. and Schnore, L. F. (eds), *The study of urbanization* (New York), pp. 213–63.

Smailes, A. E. (1947) The analysis and delimitation of urban fields, *Geography* 32, 151–61.

Smith, C. T. (1951) The movement of population in England and Wales in 1851 and 1861, *Geogrl J.* 117, 200–10.

Smith, W. (1953) *An economic geography of Great Britain* (London, 2nd ed.).

Starr, R. B. (1878) *Starr-Bowkett Societies as they are* (London).

Steindl, J. (1965) *Random processes and the growth of the firm* (London).

Stewart, J. Q. (1947) Empirical mathematical rules concerning the distribution and equilibrium of population, *Geogrl Rev.* 37, 461–76.

Stewart, J. Q. (1958) The size and spacing of cities, *Geogrl Rev.* 48, 222–45.

Stewart, J. Q. and Warntz, W. (1958) Physics of population distribution, *J. reg. Sci.* 1, 99–123.

Svennilson, I. (1954) *Growth and stagnation in the European economy* (Geneva).

Tanner, J. C. (1961) Factors affecting the amount of travel, *Road Res. Tech. Pap.* No. 51, Dept. Sci. Ind. Res., London.

Taylor, A. J. (1951) The taking of the Census, 1801–1951, *Br. Med. J.* Vol. 1, 715–20.

Temin, P. (1966) Steam and waterpower in the early nineteenth century, *J. econ. Hist.* 26, 187–205.

Thoman, R. S. (1965) Some comments on 'The science of geography', *Prof. Geogr.* 17, 8–10.

Thomas, B. (1930) The migration of labour into the Glamorganshire coalfield, 1861–1911, *Economica* 10, 275–94.

Thomas, B. (1954) *Migration and economic growth: a study of Great Britain and the Atlantic economy* (Cambridge).

Thomas, B. (1972) *Migration and urban development: a reappraisal of British and American long cycles* (London).

Thomas, E. N. (1962) The stability of distance-population-size relationships for Iowa towns from 1900–1950, in Norborg, K. (ed.), 'I.G.U. symposium in urban geography', *Lund Stud. Geogr., Ser. B.* No. 24, 13–29.

Thomas, E. N. (1967) Additional comments on population size relationships for sets of cities, in Garrison, W. L. and Marble, D. F. (eds), 'Quantitative geography, Part I: economic and cultural topics', *NWest. Univ. Stud. Geogr.* No. 13, 167–90.

Thompson, W. R. (1965a) *A preface to urban economics* (Baltimore).

Thompson, W. R. (1965b) The future of the Detroit Metropolitan Area, in Haber, W. (ed.), *Michigan in the 1970s: an economic forecast* (Ann Arbor, Mich.).

Tisdale, H. (1942) The process of urbanization, *Social Forces* 20, 311–16.

Tobler, W. R. (1963) Geographic area and map projections, *Geogrl Rev.* 53, 59–78.

Tompkins, H. (1845) *Building societies: their formation and management* (London).

Udry, J. R. (1964) Increasing scale and spatial differentiation: new test of two theories from Shevky and Bell, *Social Forces* 42, 403–13.

Ullman, E. L. (1941) A theory of location for cities, *Am. J. Sociol.* 46, 853–64.

U.S. Bureau of the Budget (1967) *Standard Metropolitan Statistical Areas* (Washington, DC).

Usher, A. P. (1921) *An introduction to the industrial history of England* (London).

Vance, J. E. (1967) Housing the worker: determinative and contingent ties in nineteenth century Birmingham, *Econ. Geogr.* 43, 95–127.

Vance, J. E. (1971) Land assignment in the precapitalist, capitalist and post-capitalist city, *Econ. Geogr.* 47, 101–20.

Veblen, T. (1942) *Imperial Germany and the industrial revolution* (New York).

Vince, S. W. E. (1955) The rural population of England and Wales, 1801–1951, *Unpublished Ph.D Dissertation*, Univ. of London.

Von Bertalanffy, L. (1950) An outline of general system theory, *Br. J. Phil. Sci.* 1, 134–65.

Von Tunzelmann, N. (1970) Technological diffusion during the industrial revolution: the case of the Cornish pumping engine, in Hartwell, R. M. (ed.), *The industrial revolution* (Oxford) pp. 77–98.

Walsh, W. D. (1966) The diffusion of technological change in the Pennsylvania pig iron industry, 1850–1870: a summary, *J. econ. Hist.* 26, 591–4.

Ward, B. (1963) City structure and interdependence, *Pap. Reg. Sci. Ass.* 10, 207–21.

Wärneryd, O. (1968) *Interdependence in urban systems* (Göteburg).

Warnes, A. M. (1970) Early separation of homes from work-places and the urban structure of Chorley, 1780–1850, *Trans. Hist. Soc. Lancs. Chesh.* 122, 105–35.

Warntz, W. (1967) Global science and the tyranny of space, *Pap. Reg. Sci. Ass.* 19, 7–19.

Webb, H. L. (1910) *The development of the telephone in Europe* (London).

Webber, C. E. (1882) Telephonic communication, *J. Soc. Arts* 30, 607–19.

Webber, M. M. (1963) Order in diversity: community without propinquity, in Wingo, L. (ed.), *Cities and space: the future use of urban land* (Baltimore), pp. 23–54.

Webber, M. M. (1964) The urban place and the nonplace urban realm, in Webber, M. M. *et al.* (eds), *Explorations into urban structure* (Philadelphia), pp. 79–153.

Webber, M. M. (1968) The post city age, *Daedalus,* 1091–1110.

Weber, A. F. (1899) *The growth of cities in the nineteenth century: a study in statistics* (New York) (republished, Ithaca, NY, 1967).

Weber, M. (1958) *The city* (English translation by D. Martindale and G. Neuwirth) (New York).

Wells, H. G. (1902) *Anticipations* (London, 3rd ed.).

Welton, T. A. (1868–9) On population statistics, *Trans. Hist. Soc. Lancs. Chesh.* New Ser. 9, 55–92, 93–164.

Welton, T. A. (1869) On the classification of people by occupations; and on other subjects connected with population statistics of England, *Jl R. Statist. Soc.* 32, 271–87.

Welton, T. A. (1897–8) On forty years' industrial change in England and Wales, *Trans. Manchr Statist. Soc.* 153–243.

Welton, T. A. (1900) The growth of population in England and Wales and its progress in the period of ninety years from 1801–91, *Jl R. Statist. Soc.* 63, 527–89.

Welton, T. A. (1911) *England's recent progress: an investigation of the statistics of migrations, mortality, etc. in the twenty years from 1881 to 1901* (London).

Wetham, E. H. (1964) The London milk trade, 1860–1900, *Econ. Hist. Rev.* 17, 369–80.

Wilbanks, T. J. and Symanski, R. (1968) What is systems analysis? *Prof. Geogr.* 20, No. 2, 81–5.

Willatts, E. C. and Newson, M. G. C. (1953) The geographical pattern of population changes in England and Wales, 1921–51, *Geogrl J.* 119, 431–50.

Williams, T. (1969) Social gradients in the city: a trend-surface analysis of enumeration district data for inner London, *Discussion Pap.* No. 32, Dept. of Geogr., London Sch. Econ.

Williamson, J. G. (1965a) Regional inequality and the process of national development: a description of the patterns, *Econ. Dev. cult. Change* 13, 3–45.

Williamson, J. G. (1965b) Antebellum urbanization in the American northeast, *J. econ. Hist.* 25, 592–608.

Williamson, J. G. and Swanson, J. A. (1966) The growth of cities in the American north east, 1820–1870, *Explor. entrepreneurial Hist.* 4, 44–67.

Wilson, A. G. (1970) *Entropy in urban and regional modelling* (London).

Woldenberg, M. J. (1967) Concepts and applications: spatial order, *Harvard Pap. Theoretical Geogr.* No. 1, part 2.

Wrigley, E. A. (1965) Geography and population, in Chorley, R J. and Haggett, P. (eds), *Frontiers in geographical teaching* (London), pp. 62–80.

Wrigley, E. A. (1967) A simple model of London's importance in changing English society and economy, 1650–1750, *Past and Present* 37, 44–70.

Yeates, M. H. (1965) Some factors affecting the spatial distribution of Chicago land values, 1910–60, *Econ. Geogr.* 41, 57–70.

Zipf, G. K. (1941) *National unity and disunity* (Bloomington, Indiana).

Zipf, G. K. (1949) *Human behavior and the principle of least effort* (Cambridge, Mass.).

Author index

General index

For Product Safety Concerns and Information please contact our EU representative GPSR@taylorandfrancis.com
Taylor & Francis Verlag GmbH, Kaufingerstraße 24, 80331 München, Germany

www.ingramcontent.com/pod-product-compliance
Ingram Content Group UK Ltd.
Pitfield, Milton Keynes, MK11 3LW, UK
UKHW010957180926
475701UK00004B/409
9780415851862